Unity and Time in Metaphysics

Unity and Time in Metaphysics

Herausgegeben von

Ludger Honnefelder
Edmund Runggaldier
Benedikt Schick

Walter de Gruyter · Berlin · New York

♾ Gedruckt auf säurefreiem Papier
das die US-ANSI-Norm über Haltbarkeit erfüllt.

ISBN 978-3-11-021234-1

Library of Congress Cataloging-in-Publication Data

Unity and time in metaphysics / edited by Ludger Honnefelder, Edmund Runggaldier, and Benedikt Schick.
p. cm.
Proceedings of a conference held Sept. 26–28, 2007 in Berlin, Germany.
Includes bibliographical references and index.
ISBN 978-3-11-021234-1 (hardcover : alk. paper)
1. Whole and Parts (Philosophy) – Congresses. 2. Time – Philosophy – Congresses. 3. Change – Congresses. I. Honnefelder, Ludger. II. Runggaldier, Edmund. III. Schick, Benedikt.
BD396.U55 2009
111'.82–dc22
2009016237

Bibliografische Information der Deutschen Nationalbibliothek

Die Deutsche Nationalbibliothek verzeichnet diese Publikation in der Deutschen Nationalbibliografie; detaillierte bibliografische Daten sind im Internet über http://dnb.d-nb.de abrufbar.

Printed in Germany

Druck und buchbinderische Verarbeitung: Hubert & Co. GmbH & Co. KG, Göttingen

PREFACE

Unity is a fundamental problem in ontology. The way we think about unity forms and distinguishes our ontological positions, and it has an influence on many other problems. Assumptions about unity have implications for both ethical and anthropological theories.

Unity can be explored from several different perspectives. Especially the questions concerning the relation of time and unity are controversial. What is temporal unity? Is there unity through time? What is the relation between perdurantism and endurantism? Is it reasonable to argue for the assumption that persons remain the same throughout their lives? What are the roles of potentiality, powers and dispositions in accounting for the temporal unity of things?

These questions were discussed in the course of the international conference 'Unity and Time as Problem in Metaphysics: Persistence and Individuality'. The conference took place in Berlin from September 26th to September 28th, 2007. Ludger Honnefelder, Christof Rapp, Edmund Runggaldier and Jan Szaif organized the conference. The papers published in this book were given during the conference.

We are obliged to thank the *Guardini Stiftung* and the *Deutsche Forschungsgemeinschaft* for their generous support; we also thank the Humboldt University's Department for Theology for their hospitality and assistance.

Ludger Honnefelder
Edmund Runggaldier
Benedikt Schick

CONTENTS

Lynne Rudder Baker:
Identity Across Time: A Defense of Three-Dimensionalism 1

Ludger Jansen:
Unity and Constitution of Social Entities 15

Uwe Meixner:
The Non-Physicalness of Material Objects 46

E. Jonathan Lowe:
Serious Endurantism and the Strong Unity of Human Persons 67

Cody Gilmore:
Why Parthood Might Be a Four-Place Relation,
and How It Behaves If It Is 83

Frank Hofmann:
An Alternative to Endurantism and Perdurantism:
Doing without Occupants 134

Jan Szaif:
The Beginnings of Greek Ontology
and the Question of Temporal Being 151

Simo Knuuttila:
New Ideas on Subject and Identity in Medieval Logic 183

Ludger Honnefelder:
Time and Existence 198

Edmund Runggaldier:
Against Conventionalism: Powers and Functions 210

Stephen Mumford:
Powers and Persistence 223

Godehard Brüntrup:
Natural Individuals and Intrinsic Properties 237

List of Authors 253

IDENTITY ACROSS TIME: A DEFENSE OF THREE-DIMENSIONALISM

LYNNE RUDDER BAKER

Abstract

To determine how a material object exists throughout an extended period of time, it is useful to ask: How does an object undergo change and still survive? Both three-dimensionalists and four-dimensionalists can answer this question. Three- and four-dimensionalists differ, however, in key claims about time and existence. I defend three-dimensionalism, first by countering two four-dimensionalist objections to three-dimensionalism, then by considering untoward consequences of four-dimensionalism that three-dimensionalism avoids. Four-dimensionalism has an anemic conception of material objects: (i) it cannot account for the significance of ordinary objects' going out of existence altogether; (ii) it makes it difficult to understand ourselves; and (iii) it seems to clash with presuppositions of morality.

The question of persistence is the question of how a material object exists throughout an extended span of time. The question becomes vivid if we ask how a single object can undergo change and still survive. Prima facie, an object changes if it has different and incompatible properties at different times: a fence that is all-white is painted and later is all-green. But nothing can be both all-white and all-green. So, how can we understand the persistence of an object (the fence, say) through change?

There are broadly two ways to understand persistence and change: three-dimensionalism and four-dimensionalism. According to three-dimensionalism, every material object has three spatial dimensions, and persists by enduring through time; the whole three-dimensional object exists at different times.[1] A 3D object has spatial parts, but no temporal parts. Trivially, a three-dimensionalist may say, for example, that your adulthood is a temporal part of your life. But in contrast to a four-dimensionalist, a three-dimensionalist does not regard your adulthood as itself any sort of object. Your adulthood is part of your life or history or career, but it is not a part of the entity that is you. Your adulthood may be represented as an ordered pair of you and a temporal interval; there is no unique object that is you-during-that-interval. There is just you and times at which you exist. The career, or life, or history of a 3D object should not be conflated with the object itself. Although being in Berlin now is part of my career or life or history, if three-dimensionalism is correct, when I am in Berlin, literally all of me is in Berlin, not just a temporal part of me.[2]

According to four-dimensionalism, every material object has four spatiotemporal dimensions, and persists by "perduring" – having a series of four-dimensional temporal parts. Temporal parts are thought of as analogous to spatial parts. On the 4D view, just as there's a part of you at each spatial sub-region of the spatial region that you occupy, there's a part of you at each temporal sub-interval of the temporal region that you occupy. You are extended in time just as you are in space. What is present at any moment during an object's existence is only a temporal part of the whole object. A temporally-extended temporal part (like your adulthood) is a sum of instantaneous temporal parts each of which exists only for an instant. According to four-dimensionalism, if this microphone exists from t_1 to t_5, at each moment that it exists there is a different temporal object – the microphone-at-t_3, say. The whole microphone (the 4D entity or worm) is the mereological sum of the instantaneous parts. According to four-dimensionalism, neither you nor I is ever all in one place; only a part

1 It does not follow that all its properties or parts exist whenever the 3D object exists; as we shall see, a 3D object's parts and properties – as well as the 3D object itself – are temporally indexed: Your appendix is part of you at one time but not at another time. Nor is three-dimensionalism committed to presentism as a theory of time and existence. (See Haslanger, 323-326.) According to presentism, ontology is constantly changing. Only what exists currently is real. There is not even a prima facie conflict between three-dimensionalism and special relativity theory if three-dimensionalists reject – as I do – presentism. "There is an unproblematic sense in which an enduring object can occupy a four-dimensional volume in space-time without being itself four-dimensional." Balashov, p. 452.

2 Standardly, three-dimensionalism is characterized as an object's being "wholly present" at moments of time. Recently, Thomas Hofweber and J. David Velleman have argued that this is incoherent. "If we can conceive of an object's extent as divisible into sub-extents – into sub-regions of space or sub-intervals of time – then we can of the object itself as divisible into parts filling those sub-extents." Hofweber and Velleman, p. 2. Although I think that the objection (like four-dimensionalism generally) conflates an object with its career or life or history, I avoid the term "wholly present". If one takes temporal parts to be temporal extents, then temporal parts should not be treated as objects. But I cannot argue for that here.

of you is in Berlin. You have different temporal parts at different places and times.[3]

The relation between the temporal parts or stages of an object is not strict identity. Strict identity is classical identity, necessary identity. If a and b are strictly identical, then a cannot exist unless b does and conversely; a and b cannot differ in any of their properties–including temporal and modal properties. If a is necessarily rational, so is b and conversely. If a and b are strictly identical, then a and b are one and the same object.[4] We also use the term 'identity' for weaker relations – contingent identity, relative identity, temporal identity, and qualitative indiscernibility – but these relations are not strict identity.

If four-dimensionalism is true, then there is no strict identity across time. If x and y are strictly identical, then x and y cannot differ in any of their properties. Since the all-white fence and the all-green fence have incompatible properties – one is all-white, the other is all-green – it seems that the all-white fence is not strictly identical to the all-green fence. But if they are not strictly identical, how do we account for the change effected by the painting of the fence? The four-dimensionalist has a ready answer: the fence is a spacetime worm that is a succession of (spatio-) temporal parts, one of which is all-white and another of which is all-green. The all-white temporal part of the fence is a distinct object from the all-green temporal part of the fence. And both of those temporal parts are distinct from the very short-lived temporal part in which exactly one half of the fence is white and the other half is green.

In order to solve the problem of change, the four dimensionalist must take the proper subjects of properties to be the temporal parts, not the fence as a whole. The fence changes color in virtue of having different temporal parts of different colors: the fence-at-t_1 is all-white, and the fence-at-t_2 is all green. The temporal parts undergo no change; they simply succeed one another. So, manifest change is explained in terms of entities that do not undergo any change. Persistence, according to four-dimensionalism, is not a matter of strict identity over time; rather 4-D persistence is a matter of being related by the some other relation such as *being a temporal part of the same 4D entity as*,[5] where a 4D entity is either a

3 Sider, p. 58. Four-dimensionalists take an atemporal parthood relation to be primitive, but can define 'part-at-t', which three-dimensionalists take to be primitive.

4 The object a and the object b are numerically identical if and only if "they" are one thing. So, if a and b are strictly identical, they are numerically identical. But, on my Constitution View, the converse does not hold. If the object a constitutes the object b at t, they are numerically identical at t without being strictly identical. You and the body that constitutes you now are numerically identical now without being strictly identical. See Baker 2007a.

5 Balashov, p. 451. Other names for this relation include "genidentity" (Carnap), the "unity relation" (John Perry), the "I-relation" (David Lewis). See Sider, p. 202.

temporal part or a mereological sum of temporal parts. Any matter-filled region of spacetime – no matter how gerrymandered or disconnected – is a four-dimensional object; so, every 4D bit of matter is a temporal part of uncountably many 4D entities.[6]

Taking David Lewis and Theodore Sider as paradigmatic four-dimensionalists, four-dimensionalism is committed to a radical conventionalism about the objects that we interact with. Ontologically speaking, all that fundamentally exist are instantaneous temporal slices of elementary physical particles and their sums. Ordinary objects are just sums of temporal parts that we choose to name: The temporal parts that I label 'this-microphone-at-t_2' and 'this-microphone-at-t_4' are parts of a single 4D entity solely because of how we choose to use the word 'microphone'. Whether or not a particular sum of temporal parts is a microphone or a person or any other ordinary object depends on our semantic decisions.[7] Some four-dimensionalists may want to avoid radical conventionalism of ordinary objects by taking the whole worm to be more fundamental than its temporal parts; but on that approach, the notion of temporal parts presupposes persistence and change and hence cannot explain them. Since on this weaker conception, temporal parts cannot explain persistence and change, I'll stick to the robust Lewis-Sider view.[8]

To sum up, a robust four-dimensionalist describes manifest change – as in the color of the fence – in terms of entities that do not themselves undergo change. A four-dimensionalist describes persistence through time in terms of series of temporal parts (each of which is a distinct object), not in terms of strict identity. For the four-dimensionalist, strict identity resides in the unchanging temporal parts; but the whole fence has different temporal parts at different times, and hence is not strictly identical across

6 Sider, p, 120.

7 There is an alternative to the worm view of four-dimensionalism: the stage view. According to the stage view, ordinary predicates and proper names (like 'computer' or 'person') refer to instantaneous stages rather than to whole worms (sums of instantaneous stages). The worm and stage views agree ontologically. Both the worm and stage views hold that what ultimately exist are instantaneous temporal parts or stages. On the stage view, no person or other ordinary object lasts more than an instant, but has counterparts at other instants. Sider, p. 193. (For another stage theory, see Hawley.) The difference between the worm and the stage views is just a matter of what we choose to call a person or a computer, or whatever; one view takes ordinary terms like 'computer' or 'person' to refer to worms, while the other takes those terms to refer to stages. That is to say, the *only* difference between the worm and stage views is *semantic*: Both views suppose that what is a computer or any other ordinary object is a matter of semantic decision that makes no ontological difference. The ontology is the same for both the worm and stage views: What exists ultimately are instantaneous temporal parts or stages, and mereological sums of temporal parts. For convenience, I focus on the worm view.

8 For a weaker notion of temporal parts that presupposes persistence (and hence does not explain it), see Shoemaker, p. 255.

time. By contrast, a three-dimensionalist takes the whole 3D object to be fundamental, and describes change in terms of exemplifying different properties at different times (as in being white at one time and green at another), and different parts (as in having a rail replaced) – without appeal to entities that do not change. And the changing 3D entities are strictly identical over time (as I'll explain), despite their change of properties, of parts and of what constitute them.

In this paper, I am going to defend three-dimensionalism and strict identity across time. First, I'll sketch my Constitution View, a 3-D account of unity at a time. Next, I'll outline a view of time and existence that takes exemplification of properties to be time-indexed, and hence allows enduring 3-D objects to undergo change of properties. After mentioning two four-dimensionalist objections that beg the question against three-dimensionalism, I'll provide reasons to prefer three-dimensionalism to four-dimensionalism.

The Constitution View: A 3-D Account of Synchronic Unity

I distinguish between unity at a time and unity over time. I explain unity at a time in terms, not of identity, but of constitution, and I explain unity across time in terms of identity. For several years, I have been working on a nonreductionistic, 3-D account of unity that allows both for change and for strict identity over time. I call the view the 'Constitution View.' The idea behind the Constitution View, reminiscent of Aristotle, is that entities that we encounter are of various primary kinds essentially, and that things of different primary kinds have different persistence conditions and different causal powers. An object's persistence conditions are the limits of the changes that the object can survive, and the persistence conditions for an object depend on the object's primary kind. Primary kinds are ordered hierarchically, so that things of one primary kind in certain circumstances constitute things of a higher-primary kind. (The ordering is only partial.) For example, when pieces of wood are in certain circumstances, a new entity – a chair, say – comes into existence.

According to the Constitution View, everything that we encounter in the natural world is constituted by 'lower-level' entities in particular circumstances. The range of possible constitution relations of a thing is also determined by its primary kind. A screwdriver can be constituted by an aggregate or sum of pieces of metal and plastic, or many other things; but a screwdriver cannot be constituted by a leaf. A speck of marble dust is constituted by an aggregate of calcium, carbon, and oxygen atoms; but only when such an aggregate is in circumstances of chemical bonding is there a speck of marble dust. The speck of marble dust is not identical to its constituting sum of atoms since the marble dust would fail to exist if the relevant atoms existed in different places. So, constitution is not identity. Indeed, as we have seen, identity is a necessary relation; but constitution is not. An aggregate of calcium, carbon and oxygen atoms may constitute a speck of marble dust at one time but not another; a piece of paper may constitute a 20-Euro note at one time but not another; a human body may constitute a person at one time but not another.

Although constitution is not identity, it is not separate existence either. Constitution is a relation of unity without identity. A speck of marble dust is a fundamentally different kind of thing from an aggregate of calcium, carbon and oxygen atoms, but there is a unity between the speck of dust and the atoms that constitute it; a person is a fundamentally different kind of thing from the body that constitutes her at a certain time, but there is a unity between the person and her body. The unity of constitution is indicated by the "sharing" of property instances. Objects related by constitution can share properties by a property derivatively (dependent on its constitution relations). The piece of paper that constitutes a 20-Euro bill at t has the property of being rectangular at t nonderivatively, but of being worth more than 20 USD at t derivatively; the 20-Euro bill that the piece of paper constitutes at t has the property of being rectangular at t derivatively, and of being worth more than 20 USD at t nonderivatively. I have the property of being a person now nonderivatively; indeed, I have the property of being a person at every moment that I exist since *person* is my primary kind. But the body that constitutes me now has the property of being a person derivatively now while it constitutes something that is a person nonderivatively.[9] (This is just an overview of a theory that is worked out in great detail in Baker 2007a.)

Associated with each primary kind are persistence conditions– conditions under which objects of that kind can exist and cease to exist. *Human organism* is a primary kind, and *person* is a primary kind. Human organisms have third-person persistence conditions that depend on biological functioning; human persons have first-person persistence conditions that depend on first-person perspectives. So, you and the organism that constitutes you now can come into existence and go out of existence at different times. But you are strictly identical to yourself throughout the whole time that you exist, and your constituting organism is strictly identical to itself throughout the whole time that it exists. It is possible that you

9 Not all properties may be had derivatively. Certain classes of properties – properties expressed by terms like 'essentially', 'possibly' 'necessarily'; properties expressed by 'is identical to' or 'constitutes'; properties rooted outside the times at which it is had; properties that are a hybrid of two primary kind properties, e.g., being a human person or a cloth flag.

come to be constituted by something different from the organism that constitutes you now (say, a prosthetic body or a resurrection body); in that case, you would still be strictly identical to yourself then.[10] You would just be constituted differently.

The Constitution View allows for change of properties by indexing properties' exemplifications to times–either by taking the property itself to be relative to time (e.g., 'being 5-feet-tall-at-t') or by taking exemplification to be relative to time (e.g., 'exemplifying-at-t being 5 feet tall'. Thus, the Constitution View delivers a conception of strict identity across time–the very same entity exists at different times–and it allows for change of properties, change of parts, and change of constitution relations.

In sum, material objects are of various primary kinds, and they are constituted by objects of other primary kinds, or of sums of objects of other primary kinds. Depending on the sorts of primary kinds involved, there may be different constituters of a self-identical material object at different times. Although persistence conditions are vague, associated with each primary kind are persistence conditions that allow objects of that primary kind to undergo some range of changes and survive. So, constitution accounts for the unity of a material object at a time, and identity accounts for its unity across time.

Time and Existence

Now, I would like to sketch a 3-D view of time and existence that supports the Constitution View. My idea is this: There are two distinct basic ways of existing–in time (like you, me, and the all-white fence) and not in time (like numbers or perhaps God).[11] Existing-at-a-time is the mode of existence that we are most familiar with. Corresponding to the two ways of existing are two ways that properties may be exemplified–temporally or nontemporally. I call this 'the Bimodal View' because it recognizes two fundamental modes of existence and of property exemplification: temporal and nontemporal. All 3-D objects are temporal objects. The fundamental mode of existence for a 3-D object is existence at a time: a 3-D

10 In Baker 2007b, I argue that the Constitution View is superior to the biological view with respect to the metaphysics of resurrection.

11 Although there are two kinds of objects in the domain of the unrestricted existential quantifier – temporal objects and nontemporal objects – there are not two senses of 'exist'. There are, rather, two modes or ways of existing. Matthews 1971, and Matthews 1972.

object exists *simpliciter* in virtue of existing at some time or other. And we who exist-at-times have properties at times.[12]

In short, according to three-dimensionalism, diachronic unity is strict identity across time, where the conditions under which an object continues to exist are determined by the object's primary kind. If we think of the world as enduring through time (as a three-dimensionalist does), then it is natural to think of existing-at-a time as a fundamental mode of existence, and it is equally natural to think of the objects that we encounter as temporal objects–objects that exist in their entirety for the whole intervals at which they exist. This picture, of course, is precisely the three-dimensionalist picture of reality.

Four-Dimensionalist Worries

Let me briefly mention two objections that four-dimensionalists have mounted against 3-D views like the Constitution View, both of which beg the question against three-dimensionalism. The first is the charge that three-dimensionalists violate Leibniz' Law. Four-dimensionalists appeal to the following version of Leibniz' Law"

$$LL_{4D}: \Box\ [x = y \rightarrow \forall F(Fx \leftrightarrow Fy)]$$

This is clearly a four-dimensionalist version of Leibniz's Law. Four-dimensional objects conform to LL_{4D}, because the only objects that enjoy strict identity are temporal parts that do not undergo change.

However, the Bimodal View of existence that I just mentioned motivates a different version of Leibniz's Law for enduring temporal objects:

$$LL_{3D}: \Box\ [x = y \rightarrow \forall F,t(Fxt \leftrightarrow Fyt)]$$

This three-dimensionalist version of Leibniz' Law is clearly the appropriate one for enduring objects, and it allows that enduring objects that undergo change enjoy strict identity over time. So, I do not think that a

12 There are two ways to understand temporal instantiation of properties – one way, roughly, is to take properties themselves to be time-indexed – as in 'The fence had the property of being all-white-at-t1 and had the property of being all-green-at-t2.' There is no conflict between being all-white-at-t1and being all-green-at-t2. Another way to understand property instantiation as temporally qualified is to take the instantiation relation itself to be time-indexed – as in 'The fence exemplified-at-t1 the property of being all-white, and exemplified-at-t2 the property of being all-green.'

three-dimensionalist need worry about violating (the appropriate version of) Leibniz's Law.

The second charge against three-dimensionalism is the so-called problem of temporary intrinsics, pressed by David Lewis. Like many other philosophers, Lewis is committed to taking intrinsic properties, like the property of *being bent*, as basic. Using the example of being bent (as opposed to being straight), his view is that we must analyze 'being bent at' a time in terms of 'being bent' *simpliciter*. Otherwise, we would have unanalyzed relations, and that, he thinks, would be intolerable. Lewis says, "[I]t is one thing to have a property, it is something else to bear some relation to it. If a relation stands between you and your properties, you are alienated from them."[13] But I ask: How could your being happy at t_1 and not at t_2 possibly alienate you from your happiness at t_1?

The motivation for holding Lewis's view on 'temporary intrinsics', I believe, stems from his other metaphysical commitments about intrinsic properties–commitments that one need not share. Indeed, I want to suggest that all alteration, all change of properties of a temporal object, is relational. There is no obvious reason why the change in the fence from being all-white to being all-green should not be relative to time. A molecular duplicate of that fence would also change from being all-white to being all-green at exactly the same time. Having a property relative to a time is a fundamental kind of relationality for temporal objects–quite unlike the relationality of my body to the wall. It seems that molecular duplicates must share some relational properties (viz., to times). All that follows is that intrinsic natures of objects cannot be captured by monadic properties. So, following a suggestion of Sally Haslanger's, I conclude that no temporary properties are monadic: "all temporary properties are relations to times."[14]

From a three-dimensionalist point of view, relations to times are ubiquitous and ineliminable. Indeed, I believe that philosophers have underestimated the scope of relational properties generally. If there really are temporal objects–objects that exist *simpliciter* only in virtue of existing at times–then it is not surprising that it is an irreducible fact about them that they have properties at some times and not at other times. The so-called problem of temporary intrinsics seems like a problem only from a point of view that rejects three-dimensionalism.

To sum up: Three-dimensionalism does not violate the three-dimensionalist version of Leibniz's Law, and, given a three-dimensionalist sketch of time and existence, the problem of temporary intrinsics just

---

13 Lewis, p. 5.

14 Haslanger, p. 330.

does not arise. In short, three-dimensionalists have an unscathed account of persistence that allows for change in enduring objects.

Consequences of Four-Dimensionalism as Reasons to Prefer Three-Dimensionalism

A main reason to prefer three-dimensionalism is that four-dimensionalism has some untoward consequences–all stemming from an anemic conception of an object. In particular, four-dimensionalists cannot find a ground in the nature of things for the unity of objects either at a time or across time, or for the significance of objects' going out of existence. Moreover, four-dimensionalism makes it difficult to understand ourselves, and, finally, seems to clash with presuppositions of morality.

First, consider the unity of objects: The 4D view offers no adequate account of the synchronic or diachronic unity of objects that we encounter. The unity of you is simply a matter of how we decide to use language and concepts. On the 4-D view, you today are an object on an ontological par with such a temporally disparate object as the sum of this microphone-today and the Eiffel Tower-100 yrs ago. Any four-dimensional space filled with matter, however disconnected, is an object–ontologically, as real as you or me. From the robust four-dimensional perspective under discussion, there is no unity in the nature of things; the appearance of unity is only a result of our interest-relative choices of which objects to recognize. Unity is just a matter of how we decide to use our concepts.

Four-dimensionalists need not deny that ordinary things like cats, rocks, microphones and people exist. What they deny is this: that they are *fundamentally* different kinds of things from each other, and that they are *fundamentally* different from arbitrary sums (like sum of my husband's eyebrows at t_1, and the Pentagon at t_2). The ontology is one of filled spacetime regions, period; and there are no fundamental differences among filled regions of spacetime, each one of which is "the total career of some object."[15]

A related feature of four-dimensionalism is its treatment of ordinary objects' going out of existence. Four-dimensionalism has the unhappy consequence that if you drop my lovely carved candle into hot water, and it melts, we may *say* that the candle went out of existence; but according to

15 Sider, p. 120. Sider's own view is a stage version (not a worm version), according to which "the objects that we typically discuss, name, quantify over, and discuss....are stages." (*Four-Dimensionalism*, pp. 190-191.) Whereas worms are temporally extended, stages are instantaneous.

four-dimensionalism, we are saying no more, ontologically speaking, than we say when we say that the one-minute temporal part of the candle at noon on Christmas day in 1998 went out of existence in one minute, to be succeeded by a different temporal part of the candle. After the melting, the candle had no more temporal parts; but the sum of particles with which four-dimensionalists identify the candle continued to have temporal parts. We just stopped calling the continued temporal parts of the sum of particles a 'candle'. I believe that when we say that a candle or anything else went out of existence, we mean something more robust than four-dimensionalism can deliver.

On the Constitution View, when the candle is melted, it goes out of existence altogether. It is not that we just stop calling the constituting sum of atoms a 'candle'. The difference between the candle's existing and not existing is an ontological difference, not just a semantic difference.

In contrast to the four-dimensionalist view, the Constitution View takes each ordinary object to be of a primary kind that is suited to be constituted by (and perhaps to constitute) objects of certain other primary kinds.[16] The unity at a time of a constituted object is secured by the constitution relation. The unity across time of a constituted object is strict identity, and is secured by the primary kind of the constituted object. The primary kind can play this role of securing identity across time, because it determines the range of changes that the object can survive. So, whereas synchronic unity is a matter of constitution, diachronic unity is identity across time.

The other consequence of four-dimensionalism that I want to mention is the difficulty of making sense of ourselves in four-dimensionalist terms, and, particularly, the difficulty of making sense of our moral experience. First, consider everyday experience. Nothing that we are familiar with or care about has any ontological significance on the 4-D picture. For example, we care about persons, but if four-dimensionalism is true, then the only thing special about persons is that we care about them. A person has no more ontological significance that that an object that consists of that chair yesterday and this lecture today.

Think of how a four-dimensionalist must understand such ordinary phenomena as our making plans and then later, perhaps much later, carrying them out. Suppose that in 2006, I decide to travel to Berlin in 2007, and then in 2007, I go on the trip exactly as I planned it. How would a four-dimensionalist understand this? Well, there's one object, me-in-2006, and a distinct object, me-in-2007, and the first object makes a plan to travel to Berlin and the second object carries out the plan. Both objects

16 Ontological simples, if there are any, are not constituted entities.

are "me" because they are connected in the way that we choose to call a person. But the whole person "me" is nothing but a series of little 'me's– like me-in-2006 and me-in-2007. If I had been a four-dimensionalist, then the me-in-2006 making the plan should have been sad that she wouldn't be around to see the plan come to fruition. There are similar kinds of awkwardness when we think of anticipation and regret from a four-dimensionalist point of view. However, if three-dimensionalism is correct, then the entity that made the plan is strictly identical to the entity that carried it out. Similarly for anticipation and regret: the exact same entity that had eagerly anticipated dinner with the Chancellor was disappointed by it; the exact same entity that made a hasty decision later regretted it. Three dimensionalism offers a much more natural way to understand ourselves across time than does four-dimensionalism.

Even more jarring are the *moral* implications of four-dimensionalism. Strict identity across time seems required for a moral life. To accept responsibility is to own up to what one has done–not just to accept responsibility for what some no-longer-existing part of one has done. Suppose that you did something reprehensible yesterday–say, you mocked a student mercilessly. Can you be held responsible today for mocking a student yesterday? *Ontologically* speaking, the entity that mocked the student yesterday is an entity that no longer exists. If you were a four-dimensionalist, you could say, "Oh, that mocker of the student was just a part of me that no longer exists. There's no point in blaming me now; that part of me is long gone."

If you apologize today for mocking the student yesterday, the only way that a four-dimensionalist can construe the apology is this: A temporal part today apologized for what a different temporal part yesterday did. There are two morally important consequences here: (1) What makes it true to say that you-today are the same person as you-yesterday is solely a matter of how we decide to use the word 'person'. We could have chosen to use the word 'person' differently; indeed, we may make different semantic decisions in the future. Nothing in the nature of things makes it right or wrong to use 'person' one way or another. (2) The other moral consequence is that in the absence of identity across time, it is *never* the case that the very same entity that committed the offense gets blamed. A four-dimensionalist may object: "You are the same person (worm) today that you were yesterday; it's just that what makes it the case that you are the same worm today is that there are distinct temporal parts yesterday and today standing in the person-relation."

But that leaves the following question unanswered: Assuming that you are morally responsible today for having mocked a student yesterday, to what does a four-dimensionalist assign responsibility, in the first instance?

The four-dimensionalist says: you are responsible in virtue of having a temporal part that is responsible. But moral properties of persons do not seem to derive from, or to be dependent on, properties of the person's parts. (If you shoot someone, we do not hold your trigger-finger responsible.) It just seems false that your responsibility for mocking the student derives from the fact that a *part* of you is responsible. And even if it made sense to say that a part of you was responsible for mocking the student, the part of you that is responsible for mocking the students is not the same part that mocked the student. Anyone who thinks that justice requires identity of the exact entity that commits the offense with the entity subsequently held responsible for it should reject four-dimensionalism.

By contrast, the three-dimensionalist can say what seems to be obviously true: Moral properties like responsibility in the first instance attach to the whole person. On the Constitution View, *person* is a primary kind, and hence being a person has ontological import. The whole human person exists from the time that a human organism acquires a rudimentary first-person perspective and comes to constitute a person, until the first-person perspective is permanently extinguished. There is no entity that is you-today; there is a single entity, you, who exist at least from birth to death. If you mocked the student yesterday, then your moral responsibility is a matter of your being strictly identical to the mocker. Again, anyone who thinks that moral accountability should be grounded in identity rather than in optional semantic decisions will prefer three-dimensionalism.

So, although both three- and four-dimensionalism have accounts of persistence and change, there are a number of reasons to prefer three-dimensionalism: Three-dimensionalism can handle both synchronic unity and identity across time, can understand going out of existence as an ontological–not just a semantic–matter, and can make sense of ourselves and of our moral experience. I do not claim that these reasons show that four-dimensionalism is false; they only point out the costs of being a four-dimensionalist. Four-dimensionalism gives us no ontological purchase on the things that we interact with and care about. By contrast, the three-dimensionalist Constitution View gives theoretical backing to what I take to be the natural way to understand reality.[17]

17 I presented this paper at the conference on Unity and Time as Problems in Metaphysics: Persistence and Individuality, at Humboldt-Universität zu Berlin, 26-28 Septermber, 2007. Thanks to participants and to Gareth B. Matthews for comments.

References

Baker, L. R. (2007a) *The Metaphysics of Everyday Life: An Essay in Practical Realism*, Cambridge: Cambridge University Press.

Baker, L. R. (2007b) 'Persons and the Metaphysics of Resurrection', *Religious Studies* 43, 333-348.

Balashov, Y. (2006) 'Persistence and Space-time: Philosophical Lessons' in: Haslanger, S. and Kurtz, R. M. (eds.), *Persistence: Contemporary Readings*, Cambridge MA: MIT Press, 451-470.

Haslanger, S. (2003) 'Persistence Through Time' in: Loux, M. J. and Zimmerman, D. W. (eds.), *The Oxford Book of Metaphysics*, Oxford: Oxford University Press, 315-354.

Hawley, K. (2001) *How Things Persist,* Oxford: Oxford University Press.

Hofweber, Th. and Velleman, D. (2007) 'How to Endure', http://www.unc.edu/~hofweber/papers/endurance.pdf. Accessed 3 October, 2007.

Lewis, D. (2002) 'Tensing the Copula', *Mind* 111, 1-13.

Matthews, G. B. (1971) 'Dualism and Solecism', *The Philosophical Review* 80, 85-95.

Matthews, G. B. (1972), 'Senses and Kinds', *The Journal of Philosophy* 69, 149-57.

Shoemaker, S. (2003) 'Identity, Properties, and Causality' in: *Identity, Cause and Mind,* Expanded Edition, Oxford: Clarendon Press, 234-260.

Sider, T. (2001) *Four-Dimensionalism: An Ontology of Persistence and of Time,* Oxford: Clarendon Press.

UNITY AND CONSTITUTION OF SOCIAL ENTITIES

LUDGER JANSEN

Abstract

Is a bank note identical with the piece of paper of which it consists? On the one hand, John Searle, in his reply to Barry Smith, suggests that they are "one and the same object" that is a social or non-social object only under certain descriptions. On the other hand, Lynne Rudder Baker puts forward the claim that bank note and paper are distinct entities that are bound together by the relation of material constitution. I suggest two possible analyses for Searle's description relativity claim, the Alternative Subject Analysis and the Predicate Modification Analysis. On both accounts his identity claim gets into serious trouble. While Baker's definition of material constitution deals well with the bank note example, it fails to account for the constitution of bearerless social entities and groups. I point out five respects in which social constitution can differ from Baker's account of material constitution and discuss compositional, institutional and interactional constitution as additional varieties of social constitution.

1. Two Puzzles about Synchronic Unity

As it often happens, unity is not only an important term for philosophy but also an ambiguous one that comes in different varieties that better are to be distinguished by the philosopher in order to avoid misconceptions and fallacious reasoning.[1] For present purposes, it is important to distinguish between diachronic and synchronic unity, i.e. between unity in time and unity at a time. Diachronic unity or unity in time corresponds to the question how many things (of a certain kind) there are between two points

1 Cf. already Aristotle, *Metaphysics* V 6 on the different meanings of "one".

of time, from t_1 to t_2? Synchronic unity, on the other hand, or unity at a time, corresponds to the question how many things there are at one certain point of time t – either of a certain kind or of different kinds? In this paper, I will primarily be concerned with synchronic unity, but questions of diachronic unity will be relevant for the discussion, too.

I will start my discussion with two puzzles concerning the synchronic unity of social entities. The first puzzle concerns the unity of groups, i.e. of such social entities that involve humans as their members. Consider a situation in which four people stand together in front of the philosophy department at noon. In this situation, how many groups are there? One? Or more? And can we tell *a priori* how many groups there are or do we need to know more about the people in question?

The second puzzle concerns the unity of 'simple' social entities that do not involve human members. Consider a philosopher – for example John Searle – holding a dollar bill in his hands: How many things are there in Searle's hand? One, a piece of paper? One, a bank note? Or two, a piece of paper and a bank note?

John Searle's own answer to this question is that there is only one thing in his hands: "In my hand I hold an object. This one and the same object is both a piece of paper and a dollar bill."[2] Searle's answer, however, is not uncontested. A less parsimonious answer has been put forward by Lynne Rudder Baker. According to Baker, there are at least two things in Searle's hand, held together by the relation of constitution: "Constitution is a fundamental relation that is ubiquitous. It is the relation that obtains [...] between pieces of paper and dollar bills."[3] According to Baker, constitution "is an asymmetric relation: If x constitutes y, then y does not constitute x".[4] But any asymmetric relation is irreflexive, and thus the dollar bill and the piece of paper that constitutes it must be distinct entities. Thus there are (at least[5]) two things in Searle's hand.

Now we have to choose between two alternatives: We can follow Searle and embrace his view of the identity of the piece of paper and the bank note, supported by a brand of description relativity with respect to social objects. Or we can follow Baker's constitution view of dollar bills and say that there are two distinct objects. To be sure, there are more options. Baker herself, for one, has supplemented her constitution view with the claim that the dollar bill and the piece of paper, though distinct

2 Searle 2003, 302.

3 Baker 2000, 27.

4 Baker 2000, 44.

5 In fact, there are many more things in his hands, because the piece of paper is itself constituted by cellulose molecules, which are, in turn, constituted by certain atoms, and so on. Hence the "at least". Cf. Baker 2007, 159: "there is constitution 'all the way down' ".

things, are still numerically the same – and thus one thing only.[6] For the time being, however, I will discuss Baker's constitution view without taking into account her view about numerical oneness, and in section 3.3 I will say why. But first I will discuss Searle's view and argue that it leads into serious difficulties. Then I will turn to Baker's constitution view and show that it copes well with the puzzle about the bank note but still leaves open a lot of questions concerning our first puzzle about the number of groups in front of the philosophy department.

2. Searle's Description Relativism

2.1 Two Intuitions and a Problem

Searle's parsimonious position that there is only one thing in his hands is clearly in tune with some deeply rooted intuitions. Obviously, the dollar bill and the paper are spatially coincident. And we need only one pick to get both the dollar bill and the paper ("In my hand I hold an object."). But Searle does more than just to appeal to these intuitions – the coincidence-intuition and the one-pick-intution. He gives us an argument for the identity of dollar bill and paper. To be sure, the proposition that Searle wants to defend is much stronger. In fact, Searle argues that the notion of a social object is "at best misleading, because it suggests that there is a class of social objects as distinct from a class of non-social objects".[7] Searle claims that there are no social objects as distinct from non-social objects. But obviously, this implies that there is no bank note distinct from the paper, and thus, in our imagined situation, there cannot be two things in Searle's hands. In Searle's eyes, one and the same object can be both a natural and social object:

> In my hand I hold an object. This one and the same object is both a piece of paper and a dollar bill. As a piece of paper it is a non-social object, as a dollar bill it is a social object. So which is it? The answer, of course, is that it is both.[8]

Searle's argument can be reconstructed as the following piece of aporetic reasoning, consisting of six premises and three intermediate conclusions:

(P1) *x* is a piece of paper.
(P2) If something is a piece of paper, it is a non-social object.

6 Cf. Baker 2007, 40-42 and 171.
7 Searle 2003, 302.
8 Searle 2003, 302. The passage in focus here is also discussed in Schmechtig 1995, who also defends the notion of a social object.

(C1) *x* is a non-social object.
(P3) *y* is a dollar bill.
(P4) If something is a dollar bill, it is a social object.
(C2) *y* is a social object.
(P5) It is the very same object that is the dollar bill and the piece of paper, i.e.: $x = y$.
(C3) The very same object is both a social and a non-social object.

Now comes the problematic bit: How exactly are the predicates "is a social object" and "is a non-social object" related to each other? It would be natural to assume (P6) – that social and non-social objects form disjoint classes:

(P6) Any non-social object is not a social object.

But from (P6) and (C3) we can derive a flat contradiction. Of course, Searle wants to avoid this contradiction. As he accepts all of the premises (P1) to (P5) and the three intermediary conclusions, it is exactly this inference step from (C3) and (P6) that Searle wants to block. Here is what Searle suggests as a solution to this problem:

> But to say that [something is both a social and a non-social object] is to say that we do not have a separate class of objects that we can identify with the notion of social object. Rather, what we have to say is that something is a social object only under certain descriptions and not others, and then we are forced to ask the crucial question, what is it that these descriptions describe?[9]

Searle suggests description relativism with respect to the social or non-social character of things.[10] According to this view, whether a thing is a social object or not crucially depends on our way to talk about it. This implies that the distinction between the social and the non-social is not a distinction between social or non-social things, but between social and non-social descriptions. To say that some predicate applies to an object only "under a certain description" is to say that this predicate does not apply to that object *per se*, but only in so far as we describe that object in a certain manner. The idea is, of course, that different descriptions allow for different predications. This implies, however, that the premises (P1)-(P4) and thus the conclusions (C1) and (C2) are not properly stated. It is not *x* or *y per se* that have social or non-social character, but only *x* or *y* under a certain description, i.e. as being described as a piece of paper or as a dollar

9 Searle 2003, 302 (direct continuation of the last quote).

10 For a possible source of inspiration cf. Anscombe 1957 and 1979. Anscombe's use of the phrase in action theory has been criticised, *inter alii*, by Goldman 1971.

bill. Thus Searle's suggestion is to rephrase the realist conclusions (C1) and (C2) as the description relativistic statements (D1) and (D2):

(D1) *x* as described as a piece of paper is a non-social object.
(D2) *y* as described as a dollar bill is a social object.
(P5) $x = y$
(D3) *y* as described as a piece of paper is a non-social object.

On the surface, the aporetic character has disappeared. As the distinction between the social and the non-social has been referred to the level of descriptions, it seems not to be an obstacle for the identity of *x* and *y*. In the following, I will argue that this is only a superficial solution and that description relativism does not in fact solve the problem.

2.2 The Alternative Subject Analysis

By way of criticism, the first thing to be said is that being a dollar bill in a situation *S* does not depend on being described as a dollar bill in *S*. In general, the existence of dollar bills does not depend on descriptions, but on the ascription of a certain status to certain things and the collective acceptance of this status by the relevant group of people.[11] But I will leave this aside in order to take under scrutiny the phrase "as described as". Of course, everything in Searle's analysis hinges on the question how this phrase is to be understood. As far as I can see, there are two competing possibilities to parse such phrases. The *Alternative Subject Analysis* considers the "as described as" phrase to be part of the grammatical subject of the sentence, while the *Predicate Modification Analysis* considers it to be part of the grammatical predicate. I will discuss these two accounts in turn.
According to the Alternative Subject Analysis, the "as described as" phrase is part of the grammatical subject. From this point of view there are two different phrases modifying the "*y*" in (D2) and (D3). Thus these two propositions consist out of two long subject phrases in combination with the contrary predicates "is a social object" and "is a non-social object"

(D2S) [*y* as described as a dollar bill] is a social object.
(D3S) [*y* as described as a piece of paper] is a non-social object.

As these two propositions are propositions containing two different subject phrases, they ascribe the two contrary predicates to different subjects,

11 This is in accordance with Searle 1995.

and thus no contradiction arises. Thus the aporetic result is actually avoided. But in order to avoid the contradiction we have made the subject phrases of (D2S) and (D3S) refer to two distinct objects! Following a suggestion of Kit Fine, such referring phrases containing words like "as described as" or "in virtue of" or "qua" are sometimes said to refer to qua-objects.[12] And the reason why Fine insists that referring phrases consisting of the same base term ("*y*") but different gloss terms ("a dollar bill", "a piece of paper") refer to different things is exactly that we can ascribe to them contrary or even contradictory predicates (like "is a social object" and "is a non-social object"), in combination with the *Principle of the Indiscernability of Identicals*, e.g. in the following formulation:

> If $x = y$, then x and y have the same properties and every predicate that can, in non-intensional contexts, be attributed to x can also be attributed to y.

By *modus tollens*, if x and y do not have the same properties, x and y are not the same thing (the *Principle of the Distinctness of Discernables*). Thus the Alternative Subject Analysis is unacceptable for Searle. For where Searle sees only one object, the Alternative Subject Analysis posits two things, i.e. two qua-objects: y as a piece of paper and y as a dollar bill.

2.3 The Predicate Modification Analysis

Thus it is likely that Searle takes refuge to the Predicate Modification Analysis.[13] According to the Predicate Modification Analysis, a phrase starting with "as described as" belongs not to the grammatical subject but to the grammatical predicate of sentences like (D2). Such phrases are predicate modifiers that take predicate phrases and yield new predicate phrases. According to this approach, (D2) and (D3) consist both of the same subject phrase "*y*" but of two distinct complex predicate phrases:

(D2P) y is [as described as a dollar bill a social object].
(D3P) y is [as described as a piece of paper a non-social object].

Thus we clearly have one and the same subject in both propositions, and therefore this account is indeed more sympathetic to the identity component of Searle's view. (D2P) and (D3P), however, contain quite strange complex predicates of the form "is a (non-)F as described as G", where

12 Cf. Fine 1982.

13 I myself think that the Predicate Modification Analysis is indeed the more appropriate approach to qua-phrases (cf. Jansen 2002, 41-43), as does Peter van Inwagen, who accusses the Alternative Subject Analysis of committing the fallacy of "adverb pasting" (van Inwagen 2000, 442 = 2001, 127-128).

"*F*" is a placeholder for "social object" and "*G*" is a placeholder for "a dollar bill" and "a piece of paper", respectively. Now there may be considerable dispute about when exactly a phrase "*x*, as described as a *G*, is an *F*" is true. But in those cases, where "F" does not require a scale reference (like "big" or "good" do), "*x*, as described as a *G*, is an *F*" implies both "*x* is an *F*" and "*x* is a *G*".[14] The former wouldn't be true without the two latter being true. Thus, as "social" is not a scaling term, if *y* is as-described-as-a-dollar-bill-a-social-object, then *y* is both a dollar bill and a social object. And if *y* is as-described-as-a-piece-of-paper-a-non-social-object, then *y* is both a piece of paper and a non-social object. Thus, according to the Predicate Modification Analysis and this implication, we end up with the very problem we started with, i.e. that *y* seems at the same time to be a social and a non-social object.

Thus Searle faces the following dilemma: Either he embraces the Alternative Subject Analysis of description relativism or the Predicate Modification Analysis. If he chooses the Alternative Subject Analysis, the diversity of discernables forces him to accept that there are two objects in the game after all, namely, according to Fine, two qua-objects. If he chooses the Predicate Modification Analysis, however, he ends up with the predicates "is a social object" and "is a non-social object" being ascribed to the object as such and thus with the very contradiction that Searle tried to avoid by introducing description relativism. Thus Searle's appeal to description relativism seems to provide no way out.

2.4 Searle's Second Argument

Searle has a second argument for rejecting the idea that there is a distinct class of social objects as opposed to non-social objects:

> Again, when I am alone in my room, that room contains at least the following 'social objects'. A citizen of the United States, an employee of the state of California, a licensed driver, and a tax payer. So how many objects are in the room? There is exactly one: me.[15]

Several things can be said regarding this argument. First, the general term "object" does not carry with it a principle of counting (nor do the general terms "thing" or "entity").[16] Thus to ask how many objects are in a room is not to ask a clear question. Thus, second, the problem Searle hints at in this passage is nothing that is peculiar to social objects: In the room to-

14 Cf. Jansen 2002, 43.

15 Searle 2003, 302.

16 Cf. Lowe 1998.

gether with Searle there are his head, his legs, and his kidneys, among a lot of other bodily parts. There are two legs, ten fingers, and thousands of hairs. Still they all form only one human organism. Like many parts can form one whole, many properties can inhere in one substrate. Suppose that there is a ball, a round thing, a red thing, a leather thing, and a thing near to me. How many things are there? It may be that all the terms in this enumeration refer to one and the same ball, which is round, red, made of leather and situated near to me. Those terms do not necessarily describe distinct objects, but may apply to the same object because of distinct features of this object. In a similar manner, one and the same natural person, e.g. John Searle, can have the social status of being a US citizen, the social status of being employed by the state of California, the social status of being a licensed driver, and the social status of being a tax payer. In this case, all of these descriptions apply to the same natural person, John Searle, because this very person is the bearer of several social features. Searle's misconception is that he sees all descriptions of objects on a par, whereas in fact there are quite distinct ontological categories involved in his example and thus these descriptions do not have to make up distinct individuals (as it would have been the case were these terms, e.g., all terms for species of substances where none is a genus of the other, like "dog", "cat", or "human being"). Just as one thing can at the same time have different properties, one thing can at the same time have different social features, and it can have more than one social status.

2.5 Identity Rejected

Searle sums up the upshot of his arguments as follows:

> There is a distinction between objects made of iron and objects not made of iron. But there is not in that way a distinction between the class of social objects and the class of non-social objects, because one and the same thing can be a social object relative to one description, and a non-social object relative to another description.[17]

I have shown that Searle's arguments are not conclusive. It is by no means clear that one and the same object can be a social object and a non-social object at the same time, depending on the descriptions one uses. In addition, description relativity is a highly artificial parlance, involving either unusual subject terms or unusual predicate terms. Moreover, because of the Indiscernability of Identicals, the kind of unity provided by synchronic identity should imply the same behaviour with regard to diachronic unity.

17 Searle 2003, 303.

Searle's identity assumption, however, leads into trouble with different persistence conditions of the allegedly identical dollar bill and the paper. Let us call the dollar bill in Searle's hand Dolly and the piece of paper in Searle's hand Piecy. Now, according to Searle's identity assumption, Dolly is the very same thing as Piecy. But Dolly could cease to exist through an act of disvalidation, e.g. through a special disvalidation stamp. Through such an act the dollar bill Dolly would cease to exist. This would, however, change nothing about the existence of Piecy: the piece of paper would still be around.

Moreover, it could well have been that Piecy came into existence, but not Dolly: It could have happened that shortly after producing Piecy, the dollar were abolished. In this case, there would be a piece of paper, but no dollar note. Piecy would exist, but not Dolly. Thus, Piecy and Dolly are not identical. In fact, this difference with regard to their actual or contrafactual persistence (which, in turn, implies a difference with regard to their modal properties) is one of the main motivating arguments behind the constitution view.

Before discussing constitution, I have to deal with an objection that could be brought forward against the rejection of identity: Let us assume for the moment that all bank notes are made of paper. Now consider the set of all pieces of paper in the world and the set of all bank notes. As I rejected identity between Dolly and Piecy in particular and between bank notes and pieces of paper in general, I am forced to judge these two sets as two totally distinct sets, as no element of the set of pieces of paper is identical with any bank note. But isn't it the case that if you have all pieces of paper in the world, you also have all bank notes? Does the rejection of identity not lead into trouble with this intuition? My answer is that the problem here roots not in the rejection of identity but in the confusion of sets with huge boxes. Sets "contain" their elements in a non-spatial way. Sets are unlike huge boxes, in that sets are not concrete but abstract things, existing outside of space and time. Were there a huge box into which we put all pieces of paper in the world, this box would in fact also contain all bank notes, because – according to our assumption – all bank notes are made of paper and thus co-located with some piece of paper. Thus the bank notes would end up in the box, not because of their identity with a piece of paper, but because of their being co-located with a piece of paper. And while identity implies co-location, this is not true the other way round: Co-location, or so the constitution view will claim, does not entail identity.

3. Baker's Constitution View

3.1 Material Constitution Defined

The retreat to description relativism is not the only response possible when faced with the above. However, the premises (P1)-(P4) are very much commonsensical and the inferences to (C1) and (C2) are based on the logically impeccable *modus ponens*. Remains the identity claim in (P5), which is indeed very much open for criticism. When confronted with (C1) and (C2), it would be natural to apply the Principle of the Indiscernability of Identicals, and reject (P5). This is exactly what the constitution view does. According to the constitution view that has been put forward by Lynne Rudder Baker, the synchronic unity between the bank note and the piece of paper is not the unity of identity but the unity of constitution. A paradigm example for the relation of constitution is the relation between a statue and the lump of stuff of which it consists. The lump of stuff is the material substratum without which the statue could not exist, but in order for the statue to exist, more is required than just a lump of stuff: Statues only exist, as Baker puts it, "in relation to an artworld",[18] i.e. in a context of social practices that consider some objects as pieces of art, as having aesthetic value, and so on. Moreover, the lump and the statue have different properties: The statue "may be defective, substandard, well or badly made, valuable, ugly, Romanesque, exchanged, insured, or admired even though the alloy which makes it up it is not."[19] Most prominently, statue and lump have different persistence conditions:

> Even if God created the statue (and, of course, the lump) *ex nihilo*, and the statue remained in existence and unchanged for a year, after which God annihilated the statue (and the lump), the lump had the property 'could survive radical deformation' and the statue did not have that property. And the statue had the property 'is necessarily conterminous with a statue', and the lump did not have this property.[20]

Any heavy deformation would destroy the statue, whereas the lump of stuff would still be a lump of stuff. Thus the persistence conditions of lump and statue also imply different modal properties: Our deliberations about the diachronic unity entail statements about the synchronic unity. Because of these arguments, constitution theorists do not conceive of the lump of stuff and the statue as being one and the same identical object.

18 Baker 2000, 34.

19 Fine 2003, 206 (italics deleted).

20 Van Inwagen 1998, 208 = 2001, 95.

The lump of stuff is the matter out of which the statue is made. Thus, the statue is materially constituted by the lump of stuff. Baker has suggested several formal definitions for this brand of constitution. Here is a slightly modified version of her most recent definition:[21]

(MC) *x materially constitutes y* at *t* if and only if there are primary kinds *F* and *G* such that at *t*:
- (a) *x* has *F* as its primary kind and *y* has *G* as its primary kind.
- (b) *x* and *y* are spatially coincident and there is no other thing that has *G* as its primary kind and is spatially coincident with *x*.
- (c) *x* is in *G*-favorable circumstances.
- (d) Necessarily, for everything that has *F* as its primary kind and is in *G*-favorable circumstances there is some spatially coincident entity that has *G* as its primary kind.
- (e) Possibly, *x* exists but no spatially coincident entity that has *G* as its primary kind.
- (f) If *x* is of one basic kind of stuff, then *y* is of the same basic kind of stuff.

Some comments are in place here. I will, in this order, elucidate (1) Baker's distinction between derivative and non-derivative properties and (2) her notion of primary kinds, then comment on (3) the definiendum and, last but not least, discuss (4) her notion of *G*-favourable circumstances:

(1) Part and parcel of Baker's view of consititution is the distinction between *derivative and nonderivative properties*. The idea is that an object *x* can have some properties nonderivatively, i.e. independently of any constitution relations it may have to other entities, be it that these entities constitute *x* or that they are constituted by *x*. If *x* is *F* nonderivatively, then – given the appropriate background – *x*'s being *F* does neither imply that *x* is constituted by something that is *F* nor that *x* constitutes something that is *F*.[22] Other objects can be *F* derivatively, i.e. by way of their constitution relations to other entities. An object *x* has a property *F upward derivatively*, if *x* has *F* because there is a constituter *c* of *x* that has *F*. And *x* has *F downward derivatively*, if *x* has *F* because *x* constitutes something that has *F*. Thus a person may have a certain weight because it is constituted by a body with this weight, and, the other way round, the body might have the right, say, to enter a cinema, because the person has a right to do so: Whi-

21 Baker 2007, 161-162. For an earlier version cf. Baker 2000, 43, 95 and 168.

22 Baker 2000, 49.

le bodies are the primary bearers of weight, persons are the primary bearers of rights, but persons can "inherit" the property of having a weight from their constituting body, and bodies can "inherit" rights from the persons which they constitute. Baker points out that the admission of both upward and downward derivation is a non-reductive feature of the constitution view.[23]

(2) The *primary kind* of a thing *x* is what corresponds to the question: "What most fundamentally is *x*?"[24] Everything has exactly one primary kind, which "goes hand in hand with its persistence conditions".[25] A thing's primary kind is something like its *infima species*, the most narrow kind it belongs to. Belonging to a primary kind is an essential property; a thing cannot exist without its primary kind: "Something that has K as its primary kind cannot lose the property of being a K without going out of existence altogether."[26] In her own versions of the definition of material consitution, Baker demands explicitly that *F* and *G* are distinct kinds.[27] This is, however, redundant, since we can infer this from clause (e). When we say that everything has exactly one primary kind, we mean to say that everything has exactly one primary kind nonderivatively. It is possible that something has other primary kinds derivatively and non-essentially. For example, while a piece of paper belongs to exactly one primary kind essentially and nonderivatively (namely the kind piece of paper), it belongs to the kind dollar note downward derivatively and hence contingently, because it is a constituter of a dollar note.[28]

(3) Baker considers her own definition as a definition of "constitution" full stop, whereas I have the more *restricted definiendum* "material constitution". This is, because I think that Baker's definition is not yet a definition of constitution "in full generality".[29] I will discuss the reasons for this in more detail in section 4, but here I can remark in passing, that according to (MC) only primary kinds – and because of clause (b) only spatial things – can have constituters or be constituters.

---

23 Cf. Baker 2000, 47. Though Baker mentions these two cases by name in this passage, she mostly deals with them in combination in the remainder of her book.

24 Baker 2000, 40; Baker 2007, 33.

25 Baker 2007, 33. Cf. Baker 2000, 39-40.

26 Baker 2007, 35. Cf. Baker 2000, 40.

27 Baker 2000, 42; Baker 2007, 161.

28 Cf. Baker 2000, 40, n. 33; Baker 2007, 34-39. Things seem to be different with upward derivation: The dollar bill belongs upward derivatively to the kind piece of paper. I take it that the laws are such that it would cease to be a dollar bill were it to consist of a different kind of matter. Thus, at least in some cases upward derivation seems to preserve essentiality.

29 Thus she describes her intention in Baker 2000, 47.

(4) A crucial phrase in Baker's definition is "*G*-favorable circumstances". This is a general term meant to cover anything that is necessary for the (material) constitution of an entity over and above the material substratum. According to Baker, there could not be statues without an art world, no persons without a first person perspective. Being embedded in an art world of artists, spectators and critiques is necessary for being a statue and thus belongs to the statue-favorable circumstances, just as, according to Baker, having a first person perspective belongs to the person-favorable circumstances a body has to be in, in order to constitute a person. And, or so we can continue, in order for there to be money, things have to be in money-favorable circumstances: They have to be embedded in social practices of selling, buying and paying, they have to be issued by the right authority, and so on.

3.2 Material Constitution Applied

But let us now put Baker's definition to work and test whether it can deal with the case of the dollar bill. Again, let "Piecy" refer to that piece of paper that Searle holds in his hands and let "Dolly" refer to that dollar bill that Searle holds in his hands. Then Piecy's primary kind is being a piece of paper, and Dolly's primary kind is being a dollar bill. It is undisputed that Piecy and Dolly are spatially coincident. Moreover, Piecy is in dollar-bill-favourable circumstances – that is, Piecy has all the properties and the origins necessary for being a dollar bill (and has not, say, been printed by a forger). And necessarily, any piece of paper that is in dollar-favourable circumstances – any piece that has such properties and origins, that is – is co-located with a dollar bill. Last but not least, it is possible that Piecy exists but no spatially coincident dollar bill, e.g. if shortly after the printing on Piecy and before issuing Dolly the dollar were abolished. In that case, Piecy would still exist, but Dolly would not.

Thus Piecy and Dolly fulfil all the requirements laid down by Baker for constitution. Thus her definition is able to account for the dollar bill Dolly being constituted by – and not being identical with – the piece of paper Piecy. And as Dolly is not identical with Piecy, Dolly and Piecy are indeed distinct entities and not, as Searle assumed, "one and the same object".

3.3 Saving Searle's Intuitions

The intuitive underpinning of the identity claim in (P5) was quite strong: Isn't it true that Searle imagines himself to hold one thing only, which is both a piece of paper and a dollar bill? Said in this way, I would totally agree. It is true to say:

(A) This piece of paper is a dollar bill.

Still, it does not follow that the piece of paper and the dollar bill are the same object. The "is" in (A) need not be the "is" of identity. Indeed, because they have different properties, it should not be the "is" of identity. Still there is an intimate connection between the piece of paper and the dollar bill: They inhabit the same region in space. This is, why we are likely to agree that there is only one thing that Searle holds in his hand, according to the intuition that no two objects occupy the same space.[30] In this wording, however, the principle is not a valid one. A bronze statue and a certain lump of bronze of necessity occupy the same space, but they are not the same object: While a heavy deformation will destroy the statue, the lump of bronze will remain a lump of bronze whatever deformation will occur. Thus, the lump of bronze and the statue are not "one and the same object". Nor is a person identical with her body, though both occupy the same space. Rather, the body is something that constitutes the person, like the bronze constitutes the statue. The same applies to the case of the dollar bill: It is not identical with the piece of paper, but the piece of paper constitutes the dollar bill, and this is, why both occupy the same space. Now, if the dollar bill and the piece of paper are not identical with each other, if they are not, as Searle invites us to accept, "one and the same object", there is no contradiction in saying that the one is a social object and the other is not.

Nevertheless, Piecy and Dolly are not totally unrelated to each other. On the one hand, they are non-identical things. On the other hand, however, Piecy is a constituter of Dolly. Thus there is a sense of synchronic unity at stake here, but it is not the unity of identity, but the unity of constitution. And this sense of synchronic unity can account for the Searlean intuitions underlying his identity assumption, but it does not run into the ontological troubles into which the identity assumption itself has lead us before. We can thus define:

30 This principle is being defended by, e.g., Burke 1994 and critically discussed in Lowe 1995a.

(UC) *x* and *y* form a unit of constitution, if and only if *x* and *y* are linked through constitution relations.

As I already remarked in section 1, Baker herself tries even more to meet the intuitions underlying the identity statement. While continuing to insist that Piecy and Dolly are distinct, she now argues for the position that Piecy and Dolly (and their ilk) while being distinct are numerically one. To advertise her position, Baker refers to "Aristotle's notion of numerical oneness without identity".[31] One may say that a seated man is a unity of a man and someone seated. If seated Socrates rises from his chair, however, that seated someone ceases to be, while the man Socrates continues to exist. The seated and the man are conceptually distinct but may indeed be numerically one man. Charlotte Witt has rightly pointed out that Aristotle's seated man – like the red round ball discussed above – is a unity composed out of a substance and accidents. Aristotle might have been counting substances – and he could say that there is numerical oneness because being seated is an accident inhering in man, and thus there is only one substance involved. Baker, however, claims numerical oneness for combinations of distinct entities belonging to different primary kinds with different persistence conditions.[32]

Questions of counting without reference to any sortal indicating a principle of counting are extremely difficult. Baker herself avows that "the 'How many' question has no application apart from some sortals".[33] And Baker goes on: "If x and y are constitutionally related, then I would deny that where x and y are, there are two things." So far, Baker's move is justified by the sortal relativity of counting: No sortal, no counting; and thus no two things. But Baker does not rest content with this. She does indeed claim: "x and y are numerically one".[34] If we take Baker at her word and strictly adhere to the need of a counting principle, she is not at all allowed to say "one" at this point: For relative to which sortal are Piecy and Dolly numerically one? Sure, they are exactly one piece of paper, because only Piecy is a piece of paper in the first place, at least nonderivatively. And they are exactly one bank note, because only Dolly is (nonderivatively) such a thing. But this is supposedly not what Baker has in mind: Also Piecy and Poundy (that English Pound note that I have in my pocket) are exactly one piece of paper – again because Poundy is not (nonderivatively) a piece of paper. What Baker has in mind seems to be something like this: Piecy and Dolly are numerically one because they make up exactly one

31 Baker 2007, 40.

32 Witt 2008.

33 Baker 2007, 171.

34 Baker 2007, 171.

unit of constitution. The sortal involved here thus seems to be "unit of constitution". But this is not true: "Unit of constitution" is not a sortal for Piecy and Dolly at all. For neither Piecy nor Dolly *are* a unit of constitution. They rather *belong* to the same unit of constitution. This is quite like saying that father and son are one and the same family. This does not make father and son numerically one. Nor does it make either of them a family. They just belong to the same family. Notwithstanding this, father and son still are numerically two, i.e. two human beings.

There is another reason not to follow Baker at this point: For any non-identical entities there is always a sortal with regard to which they are two, namely the set-theoretical sortal "element". Consider the set consisting of Dolly and Piecy. Were they numerically one, this set should have one element only. In this case, the names "Dolly" and "Piecy" would refer to the very same entity. But then the set {Dolly, Piecy} would be the same set as {Dolly}, which, in turn, would be the same set as {Piecy}. But sets are identical only if they contain the same elements. Baker, however, sustains the non-identity of Dolly and Piecy. Hence, the set {Dolly, Piecy} is to have two elements and not only one. This argument shows that, as far as absolute identity is at stake, non-identity implies numerical two-ness relative to the dummy sortal "set-theoretical element".

What, then, about our everyday parlance? If I put Dolly and Piecy into a previously empty box, it would be quite extravagant to say that there are two things inside. But then remember again the difference between sets and boxes: Not every set that contains Dolly does also contain Piecy, because they are not identical. But any box that contains Dolly will also include Piecy, because they are constitutionally related. And because Dolly and Piecy belong to the same unit of constitution, they are co-located and I need only one throw in order to put both Piecy and Dolly into the box. But that does not imply that this involves, strictly speaking, only one thing. It rather means that it involves only one *throw*. And this again does justice to the one-pick intuition.

3.4 The "Is" of Constitution

The discussion so far shows that we have to add another shade to the spectre of meanings of the word "to be": the "is" of constitution.[35] The "is" of constitution has different logical properties than, say, the "is" of identity: the constitution relation as defined by Baker is irreflexive, asymmetrical, and transitive. This can easily be shown. Clause (e) provides for irreflexivity: Nothing can constitute itself, because the primary kind of a thing cannot be both present and not present at some spatio-temporal region. And together with clause (d) it provides for asymmetry: If *x* constitutes *y*, *y* cannot constitute *x*, because if the necessitation expressed in (d) would work in both directions, (e) cannot possibly be true for any direction. And, finally, as all clauses of the definiens feature transitive characteristics, the definiendum is transitive, too.[36]

Its irreflexivity and asymmetry sharply distinguish constitution from identity. We may thus be licensed to say truly both "Dolly is Piecy" and "Piecy is Dolly", but then the "is" in these two sentences cannot possibly have the same meaning. For, as we have seen, Dolly is not identical with Piecy, thus the symmetrical relation of identity cannot be meant here. Rather, the relation between Dolly and Piecy is the asymmetrical relation of constitution. Thus, when Dolly is constituted by Piecy, Piecy is not constituted by Dolly, but rather does Piecy constitute Dolly. Thus the "is" of constitution is itself ambiguous: While "Dolly is Piecy" means that Dolly is constituted by Piecy, "Piecy is Dolly" means that Piecy constitutes Dolly.

This disambiguation of the "is" of constitution allows us to distinguish whether some property can be ascribed to a subject derivatively or nonderivatively, and if derivatively, whether it is ascribed due to upward-derivation or due to downward-derivation.[37] Again, we are licensed to say

35 Cf. Wiggins 1980, 30; Baker 2000, 54. Baker points out, that her aim is "metaphysical, not linguistic", and that she is "not postulating an ambiguity in the predicative use of 'is a person' " (2000, 54). Nevertheless, she speaks about the "is" of constitution and makes it pretty clear that she distinguishes "two ways to have a property – nonderivatively and derivatively" (2000, 55).

36 While Baker 2000, 45 argued that her constitution relation is intransitive, Zimmerman 2002 demonstrated that, to the contrary, it is transitive, and Baker 2007, 165 n. 14 now excepts this result. The point is the following: Let *x* (of primary kind *F*) constitute *y* (of primary kind *G*), and let *y* constitute *z* (of primary kind *H*). Then while it is not necessary that all circumstances that are *H*-favourable for *y* are also *H*-favourable for *x*, it is still the case that there always are *some* circumstances that are *H*-favourable for *x*. Such are, e.g., the circumstances consisting out of the conjunction of the circumstances that are *H*-favourable for *y* and the circumstances that are *G*-favourable for *x*. This is enough to guarantee the transitivity of the constitution relation.

37 Cf. Baker 2000.

both “Dolly is a dollar bill” and “Dolly is a piece of paper”. But we must be aware of the ambiguity of “is” here, too: Dolly is a dollar bill nonderivatively, but Dolly is a piece of paper derivatively. Dolly is a piece of paper because Dolly is constituted by something (namely Piecy) that is nonderivatively a piece of paper. Likewise with “Piecy is a piece of paper” and “Piecy is a dollar bill”. Piecy is nonderivatively a piece of paper, but Piecy is not nonderivatively a dollar bill. Rather, Piecy is derivatively a dollar bill, because Piecy constitutes something that is nonderivatively a dollar bill.

We can summarize these different ways of being brought about through the relation of constitution in the following table:

Dolly's ways of being	
Dolly is Piecy.	Dolly is constituted by Piecy.
Dolly is a dollar bill.	Dolly is nonderivatively a dollar bill.
Dolly is a piece of paper.	Dolly is constituted by something that is nonderivatively a piece of paper.
Piecy's ways of being	
Piecy is Dolly.	Piecy constitutes Dolly.
Piecy is a piece of paper.	Piecy is nonderivatively a piece of paper.
Piecy is a dollar bill.	Piecy constitutes something that is nonderivatively a dollar bill.

When we apply these means of disambiguation to the reductio argument from section 2, we get the following:

(P1*) Piecy is nonderivatively a piece of paper.
(P2*) If something is nonderivatively a piece of paper, it is nonderivatively a non-social object.
(C1*) Piecy is nonderivatively a non-social object.
(P3*) Dolly is nonderivatively a dollar bill.
(P4*) If something is nonderivatively a dollar bill, it is nonderivatively a social object.
(C2*) Dolly is nonderivatively a social object.
(P6*) Whatever is nonderivatively a non-social object is not nonderivatively a social object.

With these two conclusions (C1) and (C2), together with (P6*) and the Principle of the Diversity of Discernables, the constitution theorist can infer the non-identity of Dolly and Piecy. And as Dolly and Piecy are not identical, no contradiction follows. What is true, instead, is that something (i.e. Dolly) that is a social object nonderivatively is a non-social object derivatively, and that something (i.e. Piecy) that is nonderivatively a non-social object is a social object derivatively. But this is perfectly in tune with classical logic.

4. How Many Groups Are There?

4.1 Four Options

Having thus dealt with the bank note puzzle, I will now return to the group puzzle: If four people stand together in front of the philosophy department at noon, how many groups are there? To answer this question we need – implicitly or explicitly – an idea about the synchronic unity of a group. I will now discuss four possible answers to this "How many?" question, each relying on a specific account of what it is to form a group and thus on a specific account of group unity. I do not take this list of four to be exhaustive.

Firstly and most sparsely, the answer could be: There is only one group that consists of all the people present. This one group is the "maximal group", consisting out of all people present at a certain time at a certain place (and only of those). As there are four people hanging around in front of the philosophy department at noon, this maximal group consists of four members. Secondly and more affluently, the answer could be that there are many more groups: There is, of course, a group with four members, but there are also four groups with three members and six groups with two members. Thus there are, all in all, eleven groups in front of the philosophy department. In these two cases, "group" obviously means something like "(maximal) aggregate of the human beings present" or "(maximal) mereological sum of the human beings in question".[38]

Thirdly, we could respond that we cannot tell *a priori* how many groups there really are. For if we conceive of a group less formally as a relevant unit of social interaction, then we need to know more about the social interactions between these people before we can tell which of them form a group and how many groups there are: Are two of the four a loving couple? Do they all together form a group of close friends?

38 Sometimes groups are construed as sets. Cf. e.g. Rami 2005, 74 or the definition of „society" in Hawthorne 1995, 835: „A set of individuals and/or institutions in relations governed by practical interdependence, convention, and perhaps law […]." Although there is a set of the people that are in front of the department, this set itself is not in front of the department: As sets are abstract entities, they do not occupy any place in space or time at all. Groups as sets are discussed (but not endorsed) in Uzquiano 2004.

Fourthly, we could consider the possibility that there can even be more than one group consisting of the very same people.[39] The four people can - at the same time - be the faculty of a department, the advisory board of a journal, and a neighbourhood bridge club. Here, "group" means "institutional group", something that has been established by an act of institution and has some institutional structure. Again, the actual number of such groups is no matter of armchair-philosophy but requires a lot of empirical data.

4.2 Beyond Material Constitution

As different as these four options are, they share the feature that groups are somehow constituted by other entities. According to the first two options it is the people that constitute groups, even if the two options disagree about which people have to be taken into account. According to the third option, it is people plus social relations or social interactions that constitute groups. And according to the fourth option, it is people plus institutional rules or an act of institution that constitute groups.

We have seen that Baker's approach can deal with some social entities like banknotes. It can provide for the fact that some constituters (in Baker's terms, the "favourable circumstances") are extrinsic to the constituted entity. In this, it fares much better than mereological accounts of constitution, which can provide only for such constituters that are parts of the constituted entities.[40] But while Baker's brand of material constitution was sufficient to solve the bank note puzzle, it cannot at all cope with the group puzzle. There are several reasons why an account of social constitution must go beyond material constitution:

(1) Baker's definition of constitution defines a one-one relation between a single constituting entity and a single constituted entity. Social constitution, however, may be a many-one and even a many-many affair.[41] Thus, firstly, it may involve a plurality of constituters: It is a plurality of people that constitute a group, or people plus interactions, or people plus rules plus institutional acts. An obvious strategy at this point is to go for mereological sums of these as one 'single' constituter. And, indeed, Baker

39 Such cases are also discussed by Gilbert 2004 and Uzquiano 2004.

40 Mereology is discussed in Baker 2000, 179-185 and Baker 2007, 181-198. Uzquiano 2004 suggests that a group is constituted at *t* by the set of their members at *t*. But then a concrete entity (the group) would be constituted by an abstract entity (the set), which, or so it seems to me, puts things upside down.

41 The possibility of a plurality of constituters (as well as the possibility of a plurality of constituted things) is extensively discussed by Wilson 2005 and Wilson 2008.

accepts aggregates or sums as "the ultimate constituters".[42] In this way, many things can jointly constitute another thing insofar their mereological sum constitutes this thing.

It might be objected that this strategy is question-begging with regard to the mereological sums themselves, which also are groups of a kind. In the next section I will argue that sums are special in so far that they are, as I will say, 'trivially' constituted and can be dealt with easier than other groups.

(2) Secondly, social constitution may involve many constituted things, even many constituted things of the same kind: E.g., the very same people can constitute lots of groups at the same time. Baker does admit the possibility that there are several constituents of the same constituter, if only they are of different kinds.[43] In the clause (b) of her definition she explicitly demands that there is no second thing of primary kind *G* present when an *F*-thing constitutes a *G*-thing. She added this uniqueness postulate to block off certain counterexamples involving two or more persons within the same body. But why shouldn't it be possible that one organism constitutes two persons?[44] Or why shouldn't it be possible that a piece of paper constitutes two letters, each written on one of its two sides? Or even that the very same scratches of ink constitute two different letters at once – though in two different languages?[45] In case some external things are relevant for the constitution, such things seem to be totally acceptable: The ink scratches need codes like an alphabet and a language in order to constitute a letter, and there seems to be no contradiction in the assumption that one and the same pattern of ink scratches constitute different letters with respect to different codes. Thus this addition to Baker's clause (b) is not necessarily helpful if it comes to social ontology, and I will later discuss the possibility that the same people can constitute more than one group. Thus with an eye on social constitution I am not sure whether the addition of the uniqueness postulate to her definition is indeed an improvement.

(3) For Baker, constitution is a material affair. It involves material constituters. Earlier versions of Baker's definition contained a reference to immaterial entities: "If *y* is immaterial, then *x* is also immaterial." Baker later decided to replace this with her new clause (f) documented in the

---

42 Baker 2007, 181. In a note aside she considers to use plural quantificationn to solve this problem; cf. Baker 2007, 32 n. 17.

43 Cf. Baker 2007, 164: "There could be branching: The lump constitutes a plant pot, and the lump constitutes a statue. But the plant pot does not constitute a statue." Baker wants to block a counterexample put forward by Sider 2002, 46. Cf. also Wilson 2005.

44 This possibility is rejected in Baker 2007, 162, while it is defended by, e.g., Rovane 1998.

45 The letter examples are from Fine 2000.

definition given above, which she considers to be "a slightly generalized version"[46] of the older clause.[47] Both the original and the new clause (f) are intended to block off counterexamples involving, e.g., Cartesian egos[48] or 'ectoplasmatic' ghosts.[49] It is, however, not clear to me, whether immaterial things are made of any kind of stuff at all. In any case, social constitution often involves non-material constituters or at least such constituters that are not overtly material, like interaction events, individual or collective intentions, individual or collective commitments, or obligations. And many of the things constituted are non-material, too, like companies or electronic money. Thus, social constitution goes beyond the scope of Baker's definition.

(4) Closely related to this is the fact, that Baker ties constitution to spatial coincidence.[50] Co-location is at the very heart of Baker's account: It features not only in clause (b) of her definition, but also in the 'modal' clauses (d) and (e). Social constitution, however, can be a non-spatial issue, not least because some social entities are non-spatial. As Robert Wilson remarks:

> Collective social agents "are not physically bounded entities. As such, they seem unlikely candidates for satisfying the first condition of constitution: spatial coincidence. Boards of directors, trade unions, philosophy classes, families, and the welders in a factory are or can be agents of some kind, but they are not continuous, spatially bounded, physical agents, and so cannot be spatially coincident with entities that have these features."[51]

With social constitution, both constituters and things constituted may be non-spatial entities. While some groups clearly have a spatial location, some social entities haven't got one. The bridge club may be sitting in my living room, the faculty can be assembled in the seminar room, and the dollar bill is in my hand. They all have a spatial location. But companies and bank accounts do not have locations, nor do contracts or obligations. In a way, they are "quasi-abstract" entities, as Barry Smith calls them:[52] They do not extend in space, but they extend in time: They have as history, involving a moment at which they come into existence and also a moment at which they cease to exist. Troublesome cases of this kind are

46 Baker 2007, 164.

47 Baker 2007, 161.

48 Baker 2000, 43 attributes this counter-example to Anil Gupta.

49 Cf. Zimmerman 2002, 604.

50 Note that Baker now wants to construe spatial coincidence no longer as "absolute spatial coincidence", but loosely as "near spatial coincidence" in order to account for objects with vague boundaries (Baker 2007, 161; italics deleted).

51 Cf. Wilson 2005, 67.

52 Smith 2008, 37.

all those entities that have a kind of social status without having a bearer for that status, like electronic money or companies.[53]

(5) Baker restricts her account of constitution to essential properties of constituters and things constituted. According to Baker's definition, only membership to primary kinds can ground a constitution relation. Social constitution may involve external components as diverse as social relations, the look of outside observers, appropriate institutional rules or collective acceptance of these, or collective intentions.[54] These may (co-) constitute a group without being a part or a member of the group they constitute.[55] As far as the constituter is concerned, accidental and relational aspects can be accommodated within the *G*-favourable circumstances. But this means that all external constituents, which are so crucial for Baker's view, are equally hidden within the *G*-favourable circumstances.

Baker's general account is underinformative at this point. But as far as the constituting entities are concerned, that can be dealt with by spelling out what the *G*-favourable circumstances for the *G* in question are on a case-by-case basis. This is not possible as far as the constituted thing is concerned. "Fiancé", "husband" and "widow" (or, "divorcé") are social roles that, as all social entities, are in need of constitution. But according to Baker, "husband" does not denote a primary kind. [56] It is rather, one might say, a social phase sortal. A theory of social constitution must also account for social accidents, that is those social properties (like having the age of majority), social roles (like being a husband) and social relations (like being a superior to someone) that do not make up primary kinds.[57]

(6) Finally, while identity is reflexive, symmetric and transitive (or, for short, an equivalence relation), constitution as defined by Baker is, as we have seen, irreflexive, asymmetric, and transitive. Social constitution, however, seems to have non-transitive cases: Smith, Miller and Jones constitute the neighbourhood bridge club, and each of them is in turn constituted by their bodies. But, or so it seems, Smith's body is no constituter of the bridge club, and neither does the bridge club inherit all the properties of Smith's body as derivative properties: If a bridge club has a weight at all

---

53 Cf. Smith 2003a, 2003b.

54 Cf. e.g. Sartre 1943 (on outside observers), Searle 1995 (on constitutive rules and collective acceptance), Baker 2000, 24 and Wilson 2005, 51 (on external relations), Hindriks 2006 (on acceptance dependence) and Baker 2007, 11-13 (on intention dependence).

55 For the differences between parthood and membership cf. Ruben-Hillel 1985, ch. 2.

56 Cf. Baker 2000, 40: "[...] *being a husband* [...] is not a primary-kind property: A world like ours except that it lacked the institution of marriage (and hence had no husbands) would not thereby have fewer individuals in it than our world." However, such a world would contain fewer or at least different accidental entities. Cf. also Baker 2007, 34-35.

57 On the categorical variety of social entities cf. Jansen 2005.

(which can be doubted), then it is not the weight of the body of a single member, i.e. not the weight of Smith's body. The bridge club, the three members and their bodies just belong to different levels of beings. On the other hand it might be to strong to posit social constitution as a non-transitive relation, for there may be some transitive cases, especially in the case of groups considered as mereological sums. It is to these cases that I turn now.

4.3 Mere Composition

Notwithstanding the merits of Baker's approach, these six points give us good reasons for second thoughts. I will proceed in two steps. First, I will deal with those cases of groups that are only superficially of a social nature, i.e. those groups that are mere aggregates, or mereological sums, of people. In these cases we are confronted with composition, which we may either oppose to constitution proper[58] or else consider as a very weak variety of constitution that we may dub "trivial constitution". Having discussed this, I will go on and consider non-trivial cases of social constitution.

How many groups are in front of the philosophy department? Consider again the first and second answer to this question: There is one maximal group of four persons in front of the department, or there are eleven groups of two or more members, respectively. Here, I said, "group" means something like "mereological sum of human beings". There is nothing deeply social about such mereological sums, with the exception that they comprise a plurality of human beings. But the ontology of sums makes no difference as to the nature of the elements or parts: The only thing about a part of a mereological sum that matters for the ontology of sums is (beside its having parts) its very being a part. Thus sums of humans behave no different than sums of plants, cars, or stones; they are all gouverned by the same logical axioms.

If there are the parts, there is also the mereological sum of these parts. There is no explanatory gap between the existence of the parts and the existence of the sum: The existence of the parts by itself gives rise to the existence of the corresponding sum, without the need to specify any external constituters or any sum-favourable circumstances. This is the triviality involved in the "constitution" of mereological sums. It is, indeed, so trivial that the question may be asked whether it is worth to call this rela-

58 This would be Baker's choice; cf. Baker 2007, 187 ("constitution cannot be understood as mereological composition") and 181 ("Constituted objects are not identical to any sums.")

tion by the name of "constitution" at all. However, no whole is identical with any of its proper parts taken singly. And there is an important similarity between the parthood-relation and the relation of material constitution: Both feature in accounts how or why a more complex thing exists in virtue of the existence of other things – either the parts or the material constituter. And this seems to be a good reason to consider both as legitimate varieties of constitution.

Baker herself, however, is very keen to distinguish mereological composition from constitution, in order to delineate her own theory of constitution from mereological accounts of constitution.[59] For my part, I do not want to advocate a mereological account of constitution as composition *tout court*. I only want to consider the possibility that composition is a trivial variety of constitution; I do not want to claim that all cases of constitution are cases of composition. To the contrary, I agree with Baker that all those cases that she calls "constitution" are in fact not cases of mere composition.[60]

4.4 Institution and Interaction

Let us now turn to the third and fourth answers suggested above. According to these options we were not able to tell a priori how many groups there are standing in front of the department. This is a reliable indicator that these options are much less formal answers that promise actually to transfer information about the social world. As I already pointed out, according to the third option "group" refers to a relevant unity of social interaction, and "group" refers to institutional groups according to the fourth option. I will call the varieties of constitution involved in these cases "interactional constitution" and "institutional constitution".

Institutional groups are, of course, themselves relevant units of social interactions. Associations are units of co-operation and joint commitments, companies are units of employment and commerce. Thus, in fact, institutional groups are a special case of groups as relevant units of social

59 Cf. Baker 2007, 186, where she says that her prefered terminology is to say "that constitution is not composition; composition is a mereological relation, and constitution is not".

60 Another opponent of the idea of composition as constitution is David Lewis, who claims (in Lewis 1991) that composition is a many-one variety of identity: The parts just *are* the whole. But this does not only require a revisionary logic of identity as a multigrade relation "The *xs* = the *y*", but causes also serious conflict with the indiscernability of identicals in the case of, say, the statue, if one identifies the statue with all the particles of which it is composed (van Inwagen 1998). If composition is thought of as a kind of constitution, neither of these two problems arises.

interactions. For sake of simplicity, I will first turn to the special case of institutional constitution and then consider the general case.

If we want to establish an institutional group like an association or a company, the laws of our countries tell us what we have to do: How many founding members are needed to set up a charity? With what authority do you have to register? In Germany, you need at least seven people to establish a *Verein*, and you register a charity or company with the *Amtsgericht*, the local court. These are, of course, contingent facts and they differ from one legal system to another, both historically and geographically. Nevertheless, these legal facts tell us exactly what to do in order to establish an association or a company. For many institutional groups they are highly relevant for their existence.

Now many philosophers followed Aristotle in distinguishing the principles and causes of becoming from the principles and causes of being,[61] and they took great pain to point out that a thing's constituters do not belong to its causes of becoming but to its causes of being.[62] Now, if the legal process of registering an association is this association's way of coming into existence, i.e. its cause of becoming, what are its causes of being – its constituters? The legal process cannot be a constituter of an institutional group, because the process is not coexistent with the group: The group comes into being only once the process is completed and has come to its end. And the group can exist while the legal process of its establishment withers more and more into the distant past. As we can thus exclude the legal process itself from our search for the constituters, two groups of candidates remain: first the documents and records produced in the legal process, and second the rights and obligations that are established through this legal process. The legal documents, or so I will argue, are mostly only of instrumental value: They are mnemonics for and testimonials of the rights and obligations in question. Thus the only remaining candidate for the external constituter of an institutional group is the deontic structure that comes into existence through the legal process of establishing such a group. While the legal process is what brings the institutional group into existence, it is the deontic structure of rights and duties that constitutes an institutional group. It is this deontic structure that perdures once the process is over.

With this result, let us now turn to groups as relevant units of social interactions – like a loving couple or a group of close friends. Can we broaden our previous approach in such a way as to cover these cases, too? For sure, we do not legally register our friends nor do we need a legal

61 Cf. Aristotle, *Metaphysics* V 1, 1013a 18.

62 Cf., e.g., Thomas Aquinas, *Quaestiones disputatae de veritate* 2, 3 obj. 20: Constituters are *causae esse rei* or *causae essendi*, not *causae fieri*.

authority to fall in love and to start courting. And, normally, such relations do not come along with well-defined rights and obligations. Max Horkheimer and Friedrich Pollock are said to have established a contractual agreement for their friendship.[63] But normally, we do not sign contracts with our friends, and there are good reasons to belief that personal relations like friendship cannot live on the basis of a contract alone. But maybe there is something else to be found in such groups, in the storey just below the deontic network of rights and obligations: joint commitments. One way in which a group may be a relevant unit of social interactions is exactly by its members sharing a joint commitment which gives them reasons for actions according to a group intention.[64] A commitment may come about through an explicit mutual promise, i.e. a social contract *en miniature*, thus becoming a full-fledged obligation. But a commitment can also gradually build up through a series of successful co-operative actions, thus forming an implicit understanding of a joint intention to continue this series without leading to an explicit and formal act of obligation. In any case, the joint commitment comes about through social interactions of some kind.

Another way a group may be a relevant unit of social interaction is to have members that are disposed to act in an appropriate, co-operative manner, for example because they individually have the capability or tendency to react in a fitting way.[65] Again, such a capability is likely to be acquired through a process of learning or training, triggered by a series of similar situations in which each group member has the opportunity to learn the fitting co-operative reaction. Here, too, social interaction is the way to acquire these capabilities.

While in all of these cases the past interaction is the cause of the group's coming into being, past interactions no longer exist and can thus not constitute the group.[66] The entities that perdure are the commitments and the capabilities brought about through the past interactions, and these can be said to be among the group's external constituters in these cases. In a way, they are external to the group, as they are neither members nor parts of the group itself. As these groups come about through social interactions, it is not surprising that their external constituters are entities that

63 Cf. Gumnior/Ringguth 1973, 13/16. I am indebted to Michael Großheim for pointing out to me this telling anecdote.

64 Joint commitment is, of course, the central concept of Margret Gilbert's theory of plural subjects. Cf. e.g. Gilbert 1989.

65 Such cases are extensively discussed in Baltzer 1999 and Schmid 2005. I reflect on Schmid's approach in Jansen 2007.

66 If they are no constituters at all, they are, *a fortiori*, also not *intrinsic* constituters of the group.

come about through social interactions. As there is a huge variety of such groups, there is also a huge variety of possible external constituters. They range from rights and obligations via commitments to capabilities and tendencies to act in a certain way that are being shared by the individual members.

5. Towards Social Constitution

In order to account for the bank note puzzle, I rejected Searle's identity assumption and his description relativity approach to social objects and embraced a constitution view instead. I demonstrated that material constitution as defined by Baker can cope well with the bank note puzzle, but is not apt as a general account of constitution as it is found in the social realm, because social constitution goes beyond material constitution, as is shown by bearerless social entities and groups. There are at least three ways in which groups at large can be constituted: by trivial constitution (like sets and mereological sums of humans), by institutional constitution (like associations and companies), and, more generally, by interactional constitution (like a loving couple and a group of friends). Due to this richness and flexibility of social constitution, I refrain from stating a rigorous definition of social constitution, let alone of constitution in general. Even to state a number of necessary conditions is not easy beyond the usual irreflexivity and asymmetry of constitution, if they are to embrace all of material, compositional (or trivial), institutional and interactional constitution.[67]

References

Anscombe, G. E. M. (1957) *Intention*, Oxford: Blackwell.
Anscombe, G. E. M. (1979) 'Under a Description', *Nous* 13, 219-233.
Aristotle, *Metaphysica*, ed. W. Jaeger (1957), Oxford: Clarendon Press.
Baker, Lynne Rudder (2000) *Persons and Bodies. A Constitution View*, Cambridge: Cambridge University Press.
Baker, Lynne Rudder (2007) *The Metaphysics of Everyday Life*, Cambridge: Cambridge University Press.

67 Many thanks to Lynne Baker, Andris Breitling, Ingvar Johansson, Bertram Kienzle, Arto Laitinen, Pedro Schmechtig and Barry Smith, who commented on earlier versions of this paper, and also to audiences in Berlin, Krakow and Madrid for their inspiring questions.

Baltzer, Ulrich (1999) *Gemeinschaftshandeln. Ontologische Grundlagen einer Ethik sozialen Handelns*, Freiburg B.: Alber.

Burke, Michael B. (1994) 'Preserving the Principle of One Object to a Place. A Novel Account of the Relations Among Objects, Sorts, Sortals, and Persistence Conditions', *Philosophy and Phenomenological Research* 54, 591-624.

Fine, Kit (1982) 'Acts, Events and Things' in: W. Leinfellner, E. Kraemer, J. Schank (eds.), *Sprache und Ontologie. Akten des 6. Internationalen Wittgenstein-Symposiums* [1981], Wien: Hölder-Pichler-Tempsky, 97-105.

Fine, Kit (2000) 'A counter- example to Locke's thesis', *The Monist* 83, 357-361.

Fine, Kit (2003) 'The Non-Identity of a Thing and its Matter', *Mind* 112, 195-234.

Gilbert, Margaret (1989) *On Social Facts*, London/New York: Routledge.

Gilbert, Margaret (2004) 'Collective Epistemology', *Episteme* 1, 95-107.

Goldman, Alvin I. (1971) 'The Individuation of Events', *Journal of Philosophy* 68, 761-774.

Gumnior, Helmut, Rudolf Ringguth (1973) *Horkheimer. Mit Selbstzeugnissen und Bilddokumenten*, Reinbeck bei Hamburg: Rowohlt, zuerst 1973, 6th edition 1997.

Hawthorn, Geoffrey P. (1995) 'Society' in: Ted Honderich (ed.), *The Oxford Companion to Philosophy*, Oxford: Oxford University Press, 835-836.

Hindriks, Frank (2006) 'Acceptance-Dependence: A Social Kind of Response-Dependence', *Pacific Philosophical Quarterly* 87, 481-98.

Jansen, Ludger (2002) *Tun und Können. Ein systematischer Kommentar zu Aristoteles' Theorie der Vermögen im neunten Buch der Metaphysik*, Frankfurt M.: Hänsel-Hohenhausen.

Jansen, Ludger (2005) ‚Institutionen und die kategoriale Ontologie' in: Gerhard Schönrich (ed.), *Instutionen und ihre Ontologie*, Frankfurt M./Lancaster: Ontos, 45-57.

Jansen, Ludger (2007) 'Being a Community and Being in Community. On Hans Bernhard Schmid`s study of we-intentionality', *Metapyhsica* 8, 101-109.

Koepsell, David & Moss, Laurence S. (eds.) (2003) *John Searle's Ideas about Social Reality. Extensions, Criticisms and Reconstructions*, Oxford: Blackwell (repr. of *American Journal of Economics and Sociology* 62, No. 1).

Lewis, David (1991) *Parts of Classes*, Oxford: Blackwell.

Lowe, Jonathan (1995a) 'Coinciding objects: in defence of the 'standard account'', *Analysis* 55, 171-8.

Lowe, Jonathan (1995b) 'The problem of the many and the vagueness of constitution', *Analysis* 55, 179-82.

Lowe, Jonathan (1998) 'Entity, Identity, and Unity', *Erkenntnis* 48, 191-208.

Lowe, Jonathan (2006) *The Four Category Ontology*, Oxford: Clarendon Press.

Rami, Adolf (2005) 'Institutionelle Eigenschaften' in: Gerhard Schönrich (ed.), *Instutionen und ihre Ontologie*, Frankfurt M./Lancaster: Ontos, 73-119.

Rovane, Carol A. (1998) *The Bounds of Agency. An Essay in Revisionary Metaphysics*, Princeton NJ: Princeton University Press.

Ruben-Hillel, David (1985) *The Metaphysics of the Social World*, London: Routledge & Kegan Paul.

Sartre, Jean-Paul (1943) *L'être et le néant. Essai d'ontologie phénoménologique*, Paris: Gallimard.

Schmechtig, Pedro (1995) 'Konstitution und abhängige Arten. Zu einer Ontologie sozialer Objekte' in: Gerhard Schönrich (ed.), *Instutionen und ihre Ontologie*, Frankfurt M./Lancaster: Ontos, 121-169.

Schmid, Hans Bernhard (2005) *Wir-Intentionalität. Kritik des ontologischen Individualismus und Rekonstruktion der Gemeinschaft*, Freiburg B.: Alber.

Searle, John R. (1995) *The Construction of Social Reality*, New York: The Free Press.

Searle, John R. (2003) 'Reply to Barry Smith', *American Journal of Economics and Sociology* 62, 299-309; repr. in: Koepsell/Laurence (eds.) (2003), 299-309.

Sider, Theodore (2002) [Review of Lynne Rudder Baker, *Persons and Bodies*], *Journal of Philosophy* 99, 45-48.

Smith, Barry (2003a) 'The Ontology of Social Reality', *American Journal of Economics and Sociology* 62, 285-299; repr. in: Koepsell/Laurence (eds.) (2003), 285-299.

Smith, Barry (2003b) 'John Searle: From Speech Acts to Social Reality' in: Barry Smith (ed.), *John Searle* (Contemporary Philosophy in Focus), Cambridge: Cambridge University Press, 1-33.

Smith, Barry (2008) 'Searle and de Soto: The New Ontology of the Social World' in: Barry Smith, David M. Mark, Isaac Ehrlich (eds.), *The Mystery of Capital and the Construction of Social Reality*, Chicago/La Salle IL: Open Court, 35-51.

Thomas Aquinas, *Questiones Disputatae. Vol. 1: De Veritate*, ed. Raymundo Spiazzi (1949), 8th revised edition, Roma: Marietti.

Uzquiano, Gabriel (2004) 'The Supreme Court and the Supreme Court Justices: A Metaphysical Puzzle', Nous 38, 135-153.

Van Inwagen, Peter (1998) 'Composition as Identity', *Philosophical Perspectives* 8 (1994) 207-220; repr. in: van Inwagen (2001), 95-110.

Van Inwagen, Peter (2000) 'Temporal Parts and Identity across Time', *The Monist* 83, 437-459; repr. in: van Inwagen (2001), 122-143.

Van Inwagen, Peter (2001) *Ontology, Identity, and Modality. Essays in Metaphysics*, Cambridge: Cambridge University Press.

Wiggins, David (1980) *Sameness and Substance*, Oxford: Blackwell.
Wilson, Robert A. (2005) 'Persons, Social Agency, and Constitution', *Social Philosophy and Policy* 22, 49-69.
Wilson, Robert A. (2008) 'Material Constitution and the Many-Many Problem', *Canadian Journal of Philosophy* 38, 201-217.
Zimmerman, Dean (2002) '*Persons and Bodies*: Constitution Without Mereology?', *Philosophy and Phenomenological Research* 64, 599-606.

THE NON-PHYSICALNESS OF MATERIAL OBJECTS

UWE MEIXNER

Abstract

The present paper argues for the partial non-physicalness of material objects at any time of their existence – in the sense that their spatial unity is, at any time of their existence, not of a purely physical nature. The reason for this latter fact is that the place in space of a material object is, at any time of its existence, not determined on purely physical grounds. Nonetheless, the material object's (precise) place in space is determined at any time of its existence, this being so because its place in space is co-determined by human consciousness. The paper explores the consequences of this view and defends it against various objections.

Some material objects exist at some times – this seems unproblematic. But I will argue in this paper that the spatial unity of material objects *is* problematic. I will come to the conclusion *that the spatial unity of any material object is at no time of its existence of a purely physical nature.* It follows that *no material object is at any time of its existence purely physical.*

For assume that some material object is at some time of its existence purely physical. Hence this purely physical material object is at that time spatially unified (since being an existing material object entails spatial unity) and its spatial unity at that time must be of a purely physical nature (since the nature of the spatial unity of what is purely physical must itself be purely physical). But, as I have announced, it will be shown in this paper that the spatial unity of any material object is at no time of its existence of a purely physical nature.

1. The spatial unity of material objects is not of a purely physical nature

This assertion is argued for in the following way:

Assume, for *reductio (ad absurdum)*, that the spatial unity of an arbitrary material object X is at an arbitrary time t of its existence of a purely physical nature. Then this purely physical spatial unity of it requires that it be determined on purely physical grounds where X is at t and where it is not. In what follows I will argue that it is, on the contrary, *not* determined on purely physical grounds where X is at t and where it is not.

It is said to be determined *on purely physical grounds where* X *is at* t *and where it is not* (t being a moment of the object's existence) if, and only if, *the place in space at* t *of* X is determined on purely physical grounds. And this place in space is determined on purely physical grounds if, and only if, X *exactly fills* (or *exactly occupies*) – in the purely physical sense defined below (in the Central Definition) – some place in space at t.

For if X does not exactly fill any place in space at t and *one sticks to purely physical determinants*, then X's place in space at t is *not* determined (*how* could that place be determined if X does not exactly fill any place in space at t *and* one sticks to purely physical determinants?); hence, if X's place in space at t *is* determined even though X does not exactly fill any place in space at t (and it must be determined even though X does not exactly fill any place in space at t: this much is required by the spatial unity of X as an at t existing material object), then this place is not determined on purely physical grounds. Thus, if X's place in space at t is determined on purely physical grounds, then X exactly fills some place in space at t.[1] The converse of this is a matter of course.

And what does it mean that a material object *exactly fills* (*exactly occupies*) a place in space at a moment of time? Philosophers often use this relational predicate without bothering what, precisely, might be meant by it – for example, van Inwagen (1995, 33, 35, and 81), where (on page 81) the reader is asked to entertain the possibility (as antecedent of a conditional) that an immaterial ghost occupies (i.e., exactly fills) the same region of space as a human being. I am not going to leave this predicate undefined – at least not for the cases that interest us here. Let O be any material object – or, indeed, any material being –, and V a localized (coherent) volume of space (in other words, *a place in space*):

1 The structure of the argument is like this: *First*: If nonA & B, then nonC. *Hence*: If C & nonA, then non(C & B). *Hence*: If C & B, then A. ("If, then" is taken to amount to material implication.)

The Central Definition
O *exactly fills* V *at* t if, and only if, (1) there is some material being which is in O at t,[2] and (2) every material being that is at t in O is at t in V, and (3) there is *no* part of V in which there is at t no material being that is at t in O.

Now, the trouble is that there is no material object that exactly fills, in the defined sense, any localized volume of space at any time of its existence. Of course, there is no lack of material objects O, existing at a moment of time t, and localized volumes of space V, which are (1) such that there is some material being which is in O at t, and such that (2) every material being that is at t in O is at t in V. But O has at any moment of its existence some material being in it that is spatially isolated (i.e., disconnected by intervening empty space[3]) from some other material being in it at that time. This is simply a matter of contingent fact, a matter of the way the actual world is like. Therefore, the localized volume of space V will have some part in which there is at t no material being that is at t in O. Consequently, condition (3) of the above definition is not fulfilled, and therefore O does not exactly fill V at t.

Since *no* material object exactly fills any localized volume of space at any time of its existence, also the material object X of our assumption for *reductio* does not exactly fill any localized volume of space – any place in space – at time t. Therefore, the place in space at t of X is not determined on purely physical grounds, and therefore, it is not determined on purely physical grounds where X is at t and where it is not – which is precisely what had to be shown.

Clearly, the hub of this argument is the following assertion, expressing a matter of contingent fact:

The Isolation-Assertion
Any material object has at any moment of its existence some material being in it that is (at that moment) spatially isolated from some other material being in it (at that moment).

The following quotation from van Inwagen (1995, 34) states a fact – uncontroversial in physics (at least as van Inwagen presents matters) – which strongly supports the Isolation-Assertion:

2 This condition is tantamount to the condition "O exists at t" (see Section 4).

3 "Empty space" is to be taken in the sense of "space without any material being in it," that is, in the sense of "space without anything in it that can be assigned a rest-mass." Thus empty space can be full of all kinds of non-material physical beings, like photons or fields.

> "[I]t is undeniably true that, if there are any composite material objects at all, they are composed of elementary particles and that the elementary particles that compose a given material object are not in contact."

And this is a consequence of the Isolation-Assertion:

The Denial of Democritean Bodies
There are no material objects which are at some time of their existence Democritean bodies.

A material object that is a *Democritean body* at a time t of its existence would be a material object that is completely full at t; hence there would not be any lacuna of empty space,[4] no matter how small, that spatially isolates (spatially disconnects) at t some material being in the object from some other material being in it. But this would contradict the Isolation-Assertion. The denial of Democritean bodies seems unproblematic since, apparently, such bodies would have to have infinite mass, and there is no empirical evidence for material beings with infinite mass.

Another important consequence of the Isolation-Assertion is this:

The Denial of Material Atomic Objects
There are no material objects which are at some time of their existence mereological atoms.

A material object that is a *mereological atom* at some time t of its existence would be a material object such that there is at t no material being in it that differs from it; hence, a fortiori, there would be at t no material being in the object that is at t spatially isolated from some other material being in it at t. But this would contradict the Isolation-Assertion. The denial of material atomic objects seems unproblematic, too.[5] Since any material object is, as long as it exists, a *three-dimensional space-taker*, it is *in any case* – whether one accepts the Isolation-Assertion or not – hard to see how it could have, at some time of its existence, no material being in it that differs from it. After all, a three-dimensional region of space always has a three-dimensional region of space in it that differs from it. It should be noted that the Isolation-Assertion does not (in itself) exclude that there is a *material being* that has, at some time of its existence, no material being in

4 Regarding the notion of empty space, see footnote 3.

5 It is *not* denied, of course, that there are atoms in the sense of physics; for the atoms of physics are not *mereological atoms*.

it that differs from it; *however*, given the truth of the Isolation-Assertion, such a *material being* would not be a *material object*.[6]

This is the place to point out that, necessarily, every material object is a material being, but that the reverse of this is not necessarily true: possibly, some individual being that has, as long as it exists, a (non-zero) rest-mass (in short: some material being[7]) is not, as long as it exists, a three-dimensional space-taker, and thus does not fulfill a necessary condition (on conceptual grounds) for being a material object.[8] In fact, later in this paper, a hypothesis (about the actual world) will be made use of that centrally involves material beings that are not material objects. Given the distinction between *material being* and *material object*, it should also be noted that the Isolation-Assertion does not exclude that there is a material object that has, at some time of its existence, *no material object* in it that differs from it; *however*, if such an object still had *some material being* in it that differs from it, it would not be a mereological atom in the strict (true) sense.

But now, if the Isolation-Assertion is true, as it seems to be, what, then, do material objects ultimately amount to? Take some material object O_0. Since the Isolation-Assertion is true of it, it has at a given time t of its existence two material beings in it that are spatially isolated from each other, in other words: which are thus that one cannot get from one to the other without crossing empty space (space that is empty in sense of footnote 3). If they are themselves material objects, then the Isolation-Assertion is true of them, too, and they each have at t (which is also a moment at which *they* exist) another two material beings in them that are spatially isolated from each other; and if these latter material beings are in turn material objects, then the Isolation-Assertion is true of them, too, and so on. O_0 thus dissolves at every moment of its existence into ever more fine-grained simultaneously existing material objects that are spatially isolated from each other. For visualizing this schematically, just consider the first four levels of a division-pyramid that the Isolation-Assertion gives rise to:

6 According to van Inwagen (1995, 99), current physics "strongly suggests" that some elementary particles are without proper parts, i.e., that they are mereological atoms (*simples*). To my mind, the mereological status of elementary particles is far from clear, especially if they have, as long as they exist, a (non-zero) rest-mass, that is: if they are material beings.

7 Cf. the preceding footnote.

8 It is true: the term "object" is a very colorless term and is often used as a synonym for "being" or "entity." But in this paper, the term has a more specific meaning – at least if it occurs in combination with "material."

O_0
has in it at t: O_1 O_2
have in them at t: O_{11} O_{12} O_{21} O_{22}
have in them at t: O_{111} O_{112} O_{121} O_{122} O_{211} O_{212} O_{221} O_{222}
… …

Given the Isolation-Assertion, the prospects for the *spatial unity on purely physical grounds* of the material object O_0 (or any of its material sub-objects) are even bleaker than one may, perhaps, have been initially inclined to believe. Although O_0 does not exactly fill, at any moment of time, any localized volume of space (according to the Isolation-Assertion), one still may have *hoped* of being able to assign to O_0 on purely physical grounds at each moment of its existence a *spatially distributed region of space* as *its location*, where "location" or "place in space" now has a wider sense than "localized volume of space." *Such* a region is not a localized volume of space, but a *set* of spatially separated localized volumes of space. A spatially distributed region of space could be assigned to O_0 on purely physical grounds as its location at a moment t of its existence *if* O_0 were completely divided at t into material beings, spatially isolated from each other, each of which exactly fills (in the sense defined) a certain localized volume of space at t. O_0's location at t would then be the set of the localized volumes of space that are exactly filled at t by the elements in the corresponding complete division-set for O_0 at t. But here's the catch: Would not the elements in that division-set be *at t existing material objects* (seeing that they are supposed to exactly fill certain spatially separated localized volumes of space)? In fact it seems undeniable that they would be at t existing material objects, and hence the Isolation-Assertion becomes applicable *to them*, too, and the above-mentioned *hope* – of being able to assign to O_0 on purely physical grounds at each moment of its existence a *spatially distributed region of space* as *its location* – is dashed.

2. The true nature of the spatial unity of material objects

The question of what it is that material objects ultimately amount to, given the truth of the Isolation-Assertion, is still unanswered – and not just by me. However, it is consistent with that assertion to assume that each material object is at every moment of its existence a collection – not *just* a collection, but *also* a collection[9] – of precisely located *material points* each of which has unit-mass and is spatially separated from the others (it being impossible that two material points occupy the same spatial point at the same time); the number of material points in such a discrete, discontinuous collection is usually very high, though always finite (otherwise the mass of a material object – that is: the sum of the masses of the material points which go into the object – would sometimes be infinite; but it never is infinite). This speculative hypothesis – which seems to me to be the simplest one consistent with the Isolation-Assertion – stays by and large within the confines of classical physics. Its non-classical element, of course, is the idea that the material points all have the same minimal mass: 1 unit-mass.[10] The justification for assuming that material points all have the same minimal mass is obvious: material points have *no inner structure*, and they all have *no inner structure* in the same way; hence there is no good reason to assign different masses to them.

But, setting aside the non-classical element just described, is not this – by and large – *classical* hypothesis on the nature of material objects wildly unrealistic in view of modern physics, in particular, quantum physics? And is not this a very serious drawback to that hypothesis? The response to this query must be that descriptive realism, or the lack of it, is, as a matter of fact, not a relevant issue here. As I said in the beginning of this section: given the truth of the Isolation-Assertion, the question of what it is that material objects – that is, material beings which are three-dimensional space-takers – ultimately amount to is still unanswered, and modern physics (our, to date, best physics) has certainly not increased our hopes that this question will ever be answered in a way that would satisfy the cravings of the scientific realists. Given the truth of the Isolation-Assertion, the search for a *purely physical, mind-independent solution* to the problem of the spatial unity of material objects seems to me even less promising within the framework of modern physics than within the framework of (by and large) classical physics.[11] The right way to regard the hypothesis on the nature of material objects put forward above is, therefore, this: the hypothesis – true or not – states the *best* conditions compatible with the Isolation-Assertion for still achieving – in spite of bleak prospects – *a*

---

9 This caveat needs to be added in view of the *Constitution-Statement*, which follows later in the paper. The Constitution-Statement entails that a material object is at any moment of its existence a collection of material points *plus* its (the collection's) *territory* (so to speak).

10 If one measured their masses *in gram*, one would always obtain the same very small value of *m* gram; note the analogy to Planck's constant *h*.

11 Quantum physics has no place for material points (except if they are regarded as package-like concentrations of indefinitely spread-out wavelike entities and are in consequence without a precise location).

purely physical, mind-independent solution to the problem of the spatial unity of material objects. However, *even if* that hypothesis (which, compared to other hypotheses on the nature of material objects, is clear and perspicuous) turned out to be true and those best conditions really obtained, no such solution would be forthcoming – as will become completely evident below. *All the worse* for a purely physical, mind-independent solution to the problem of the spatial unity of material objects if that hypothesis turned out to be *not true* (and, indeed, *very likely* it has already turned out to be not true).

Material points are zero-dimensional material beings that exist at some times and occupy at any moment of their existence some *point in space* (a *point* in the sense of *physical geometry*).[12] It is important to keep in mind that no material point is a material *object*. The simple reason for this is that all material objects are three-dimensional material beings, whereas every material point is a zero-dimensional material being.

If material objects are at each moment of their existence collections of spatially separated material points, it is entirely clear why the Isolation-Assertion is true of them. Moreover, given this hypothesis on the nature of material objects – *the cloud-of-material-points hypothesis* – it is immediately evident that there is no place in space, no localized volume of space that any material object exactly fills at any time of its existence. If MAP(t, O) – *the material-point-set of* O *at* t – is the set of all the material points that go into making up the material object O at the moment t of its existence,[13] then there corresponds to that set the set LOC(t, O) – *the location-set of* O *at* t. This set is defined as the set of every point in space that is occupied at t by some material point in MAP(t, O).[14] But since the elements of MAP(t, O) are all spatially separated from each other at t, the set LOC(t, O) does not constitute a place in space, a localized volume of space (nor a set of localized volumes of space). There is, indeed no lack of places in space of which LOC(t, O) is a subset (that is, a part), but, clearly, each place in space of which LOC(t, O) is a subset also comprises infinitely many points in space that are not elements of LOC(t, O).

But might one not declare that LOC(t, O) – the location-set of O at t – is *the location of* O *at* t, in another wider sense of "location" or "place in space" (such that not only localized volumes of space are possible locations for material objects)? Would it not, then, be determined on purely

12 The occupation (filling) of a *point in space* by a *material point* can only be *exact occupation (exact filling)*: a material point does not reach beyond the point in space it occupies, nor does the point in space reach beyond the material point by which it is occupied.

13 Since all the elements in MAP(t, O) go into making up the material object O at the moment t of its existence, all the elements in MAP(t, O) must exist at t.

14 Necessarily, every material point in MAP(t, O) occupies *exactly one* point in space at t.

physical grounds *after all* (in a different sense than first envisaged, but nevertheless in an entirely adequate sense) where O is at t, and where it is not, giving O *after all* a spatial unity of a purely physical nature? The answer to these questions is a resounding "No." As a material object, O is a *three-dimensional object*, and therefore LOC(t, O) (which is just a finite set of disconnected points in space) is out of the question as a candidate for being *the location of* O *at* t. It is only either a localized volume of space or, second-best, a set of spatially separated localized volumes of space that can, in principle, be *the location of* O *at* t. But, given the truth of the Isolation-Assertion, neither one of these two alternatives is feasible *on purely physical grounds* (as was seen in the previous section). In what follows I will concentrate on the first alternative, since, with regard to the *spatial unity* of a material object, it is clearly preferable to the second (and also because the second alternative can hardly be feasible without the first being feasible at least for some material objects).

The place in space of O at t – the location of O at t – (which is, in the now confirmed preferred sense, a localized volume of space) is not determined on purely physical grounds. Yet, the place in space of O at t is *determined*; this much is required by the spatial unity of O. How, then, *is* the place in space of O at t determined? Not on purely physical grounds, but certainly not without *any* physical grounds. The physical grounds are provided by the location-set of O at t, that is, by LOC(t, O). The place in space of O at t, VOL(t, O), whichever localized volume of space it turns out to be in the end, properly includes LOC(t, O); moreover, VOL(t, O) is related to LOC(t, O), its subset, in the following way: all points in space in LOC(t, O) are occupied (by material points), *whereas* the remainder of the points in VOL(t, O) is unoccupied by any material being (*empty* in the sense of footnote 3). This is how far the contribution of physical grounds to the determination of the place in space of O at t goes, and no further. The rest – the selection of one single place in space as *the* place in space of O at t from an infinite set of candidate places in space – is provided by *us*, by human subjects of consciousness. Phenomenal consciousness presents us with certain objects of phenomenal intentionality: seen and felt wooden spheres, for example. Such objects, certainly, are paradigmatic material objects. Let O be a wooden sphere. The place in space of this sphere, at a time when we experience it, is determined as an objectified upshot of our experience regarding the whereabouts of the sphere's outer boundaries, that is, ultimately as an objectified upshot of our experience of visual and tactile *local resistance* of a certain – *spherical* and *wooden* – kind. A very important factor in this feat of the human mind is the ability of our consciousnesses to present something as materially continuous and coherent, forming a bounded whole, *which*, in *mindless* reality, is no such thing.

There is no reason not to generalize on the basis of this specific example. Thus, our objectified experiential findings regarding the whereabouts of the outer boundaries of a material object at a certain time are what determines a specific place in space for that material object at that time. *The* localized volume of space (namely, a certain localized solid of physical geometry) that is ipso facto specified by our objectified experiential findings regarding the whereabouts of the outer boundaries of a material object at a certain time – just is *the place in space* of that material object at that time. This is the only way in which the place in space of a material object at a certain time – *where* it is at that time, and *where* it is not – can be determined in the (actual) world; hence it is the only way in which the spatial unity of a material object can come into the world.

3. The true nature of material objects

And it's the only way in which a material object itself can come into the world. How so? We can state for any material object O that exists at a time t:

The Constitution-Statement

O is constituted at t by {MAP(t, O), VOL(t, O)},[15]

where the connection between the two elements of this pair-set is established by LOC(t, O) – that is, $\{x$: x is a point in space & $\exists y(y \in$ MAP(t, O) & y occupies x at t)} – on the followings grounds: LOC(t, O) $\subset$[16] VOL(t, O) and non$\exists x(x \in$ VOL(t, O) & $x \notin$ LOC(t, O) & $\exists y(y$ is a material being & y occupies x at t)).[17]

As specified earlier, MAP(t, O) is the material-point-set of O at t, LOC(t, O) the location-set of O at t, and VOL(t, O) the place in space of O at t: a certain localized volume of space. The relationship between MAP(t, O), LOC(t, O), and VOL(t, O) has already been described in the preceding

15 Note that O need not be constituted at time $t' \neq t$ by the same pair-set {MAP(t, O), VOL(t, O)}: {MAP(t´, O), VOL(t´, O)} may well be different from {MAP(t, O), VOL(t, O)}. Note also that O may be different from O´ although {MAP(t, O), VOL(t, O)} = {MAP(t, O´), VOL(t, O)´} (see Section 5 for more on this).

16 "$\subset$" means "proper subset."

17 How do some concepts of physics, used for describing material objects, fit into this picture? As follows:

The *mass* of O at t: the (finite) sum of the unit-masses of all the elements in MAP(t, O).

The *volume* of O at t: the quantity of the extension of VOL(t, O).

The *density* of O at t: the mass of O at t divided by the volume of O at t.

section; that description is here merely restated, using, for brevity's sake, set-theoretical symbolism. To the description, however, there is added, in the Constitution-Statement, the explicit assertion of what it is that a material object, existing at a given time, *amounts to.* (That assertion, by the way, should not be misunderstood as saying that a material object, existing at a given time, coincides at that time with a certain pair-set, an abstract object; I trust that the use of the language of set theory for *conveniently representing* – *not* for literally describing – ontological facts is readily understandable without needing long-winded explanations.)

Given basic physical reality, conceived of in the way described (involving material points in the way described), it is determined on purely physical grounds at a given moment of time t which material points exist at that time and where they are located. It is to be assumed that not every finite set of material points that exist at t is fit to constitute[18] a material object at t: there are bound to be certain restrictions that determine highly selective *criteria of inner material coherence* which must be fulfilled by any set of material points that is fit to constitute a material object at a time.[19] But one may assume that these criteria of coherence relate to matters that are purely physical in nature. Hence, given basic physical reality, it is determined on purely physical grounds at t which finite sets of material points that exist at t are *fit to constitute* a material object at t. But this fact of determination notwithstanding, it is *not* determined on purely physical grounds which objects exactly are the material objects that *are constituted* at t. Therefore, the material objects constituted at t – that is, existing at t[20] – *do not supervene* on basic physical reality. And therefore they are not purely physical objects.

For constituting a material object O at t *two things* are required (in conformance to the Constitution-Statement): (1) a finite set MAP of material points existing at t that is *fit to constitute* a material object at t; this set will become the material-point-set of O at t; and (2) a localized volume of space VOL in which all the elements in MAP are located at t and which is otherwise empty; this volume of space will become the place in space of O at t. The trouble is that the first factor of constitution does not determine the *second*; the second factor is largely (though not entirely) independent of the first. As I have argued, it is *we* who determine the second

18 More precisely (in view of the Constitution-Statement): "is fit to be one of the two factors in constituting."

19 Those criteria are not easy to specify *even though* they are not criteria of inner material coherence which, if fulfilled, are *sufficient* for the constitution (or existence) of a material object – there aren't any such criteria –, but criteria of inner material coherence which, if fulfilled, make the constitution (or existence) of a material object *possible.*

20 If a material object exists at time t, it is constituted at time t, *and conversely.*

factor in the constitution of a material object at a given time t – and if not every set of material points that is fit to constitute a material object at t *actually constitutes* (i.e., co-constitutes) a material object at t,[21] then it seems that it is also *we* who determine the first factor in the constitution of a material object at t; but I am not going to pursue this line of thought here further.

In determining the second factor of constitution, we more or less automatically observe certain implicit rules regarding the connection between MAP(t, O), LOC(t, O), and VOL(t, O), or, better, *guidelines* (since not all of these rules allow of no exception) – *guidelines* regarding the connection between MAP(t, O), LOC(t, O), and VOL(t, O) which come *in addition* to the absolutely minimal (non-negotiable) strictures specified in the Constitution-Statement itself. For example, the following guidelines – O (and later also O´) being a material object, and t a time of its existence:

Guideline 1
VOL(t, O) should fit MAP(t, O), or, more properly speaking, LOC(t, O), *very closely* (by our conscious, or more generally speaking: *cognitive*, lights!).

The surface of VOL(t, O), in enveloping LOC(t, O), should not be like a loose gown, but like a tightly fitting one. But we also observe the following guideline that relativizes (to a certain extent) Guideline 1:

Guideline 2
Although VOL(t, O) should fit LOC(t, O) very closely, VOL(t, O) should be a comparatively simple localized solid of physical geometry – unless there are reasons for eschewing simplicity.[22]

If a material object is constituted at a time – and, to repeat, it is *we* who constitute it – then all the material objects in it at that time are constituted together with it. In providing the places in space to these material objects, we observe the following guideline:

21 *If* every set of material points that is fit to constitute a material object at t actually constituted a material object at t, we would be confronted, it seems, with the so-called *problem of the many*. See Unger (1980) and, for further discussion, Lewis (1999). Perhaps this problem is a truly serious problem, perhaps not. (Why not accept that there are 1001 cats sitting on the mat, *given* that they all spatially coincide with each other to more than 99 percent?)

22 Such reasons – to give an important example – are present when living organisms are provided with places in space. What fills a living organism's intestines does not occupy space that belongs to the living organism's place in space (otherwise what fills a living organism's intestines would be a part of the living organism). Hence the place in space of the living organism is not a simple localized solid of physical geometry.

Guideline 3
If O´ is a material object that is at t in the material object O, then VOL(t, O´) $\subseteq$ VOL(t, O).

It goes without saying that these guidelines leave considerable latitude to our determination of a place in space for a material object, constituted at time t, of which the material-point-set at t is given. Other guidelines can presumably be added to those presented here, but this is not going to change the basic situation.

4. The tying up of some loose ends

Some loose ends need to be tied up. In this paper, the following mereological predicates have played an important role: "(material being) X is in the localized volume of space V at t," "(material being) X is in the material object O at t," "(localized volume of space) V´ is a part of (localized volume of space) V." All three predicates occur, for example, in the Central Definition, the definition of the predicate "O exactly fills V at t" in Section 1. Here follow the definitions of these three predicates:

D1
V´ is a part of V $=_{\text{Def}}$ V and V´ are localized volumes of space and V´ $\subseteq$ V.

D2
X is in the volume of space V at t $=_{\text{Def}}$ X is an at time t existing material being, V a localized volume of space, and LOC(t, X) $\subseteq$ V.

D3
X is in [*or*: a part of] the material object O at t $=_{\text{Def}}$ X is an at time t existing material being, O an at t existing material object, and LOC(t, X) $\subseteq$ LOC(t, O).[23]

Two remarks: (1) We have already defined the location-set of a *material object* O at a time. The more general concept of the location-set of a *material being* X at a time t, LOC(t, X), which is employed in the above definitions, is defined in exactly the same way: as $\{x$: x is a point in space & $\exists y$(y

23 According to this definition, the phrases "O is an at time t existing material object" and "some material being is in the material object O at time t" are (analytically, or broadly logically) equivalent.

$\in$ MAP(t, X) & *y* occupies *x* a t)}, *assuming* material-point-sets for material beings in general, and not only for material objects. (2) Since material points must be counted among the *material beings* (though not among the *material objects*), the above definitions of *in-being* apply also to them.[24]

The important concept of *material part-object* is defined as follows:

D4

O´ is at t a material part-object of O $=_{\mathrm{Def}}$ O´ and O are at time t existing material objects, and VOL(t, O´) $\subseteq$ VOL(t, O).

It follows from the definiens of this definition that the material-point-set of O´ is included in the material-point-set of O: MAP(t, O´) $\subseteq$ MAP(t, O), and hence that the location-set of O´ is included in the location-set of O: LOC(t, O´) $\subseteq$ LOC(t, O). This is exactly as it should be.

Suppose that O´ and O are at time t existing material objects and VOL(t, O´) $\subseteq$ VOL(t, O), but that some material point p that is an element of MAP(t, O´) (and hence exists at t) is not an element of MAP(t, O). Consider the point in space that is occupied at t by p, s(p). s(p) is an element of LOC(t, O´), but it is not an element of LOC(t, O) (otherwise p would be an element of MAP(t, O),[25] contradicting the assumption). Since s(p) is an element of LOC(t, O´), it is an element of VOL(t, O´), because LOC(t, O´) $\subset$ VOL(t, O´) (according to the Constitution-Statement applied to O´). Hence s(p) is an element of VOL(t, O) (according to the assumption, VOL(t, O´) $\subseteq$ VOL(t, O)). But according to the Constitution-Statement applied to O: non$\exists x$($x \in$ VOL(t, O) & $x \notin$ LOC(t, O) & $\exists y$(y is a material being & y occupies x at t)). Therefore, because of s(p) $\in$ VOL(t, O) & s(p) $\notin$ LOC(t, O), we finally get: non$\exists y$(y is a material being & y occupies s(p) at t). But this is false, since p, which is a material being, occupies s(p) at t.

In view of the result just reached and of Guideline 3, it is apparent that, for material objects O´ and O, the phrases "O´ is in the material object O at t" (*as interpreted by* D3) and "O´ is at t a material part-object of O" (*as interpreted by* D4) are equivalent statements, which, again, is exactly as it should be.

24 Note that the material-point-set of a material point p at a time t when it exists is {p}; LOC(t, p), therefore, turns out to be {x: x is a point in space & p occupies x a t}, and this set is identical with the set {the point in space occupied by p at t}.

25 Remember: it is impossible that two material points occupy the same spatial point at the same time.

5. The identity of material objects

One of the most time-honored principles of metaphysics is the following:

The Location-Principle for the Identity of Material Objects
If O´ and O are both material objects that both exist at time t and the place in space of O´ at t is identical with the place in space of O at t, then O´ is identical with O.

There are certain well-known objections to this principle. Tib is defined as being the cat Tibbles without its tail, and one day Tibbles' tail is destroyed (while Tibbles continues to exist). Then, at time t after that day, the place in space of Tib at t is identical with the place in space of Tibbles at t, and yet Tib and Tibbles are *non-identical* at t existing material objects – because they do not have the same properties, as one notices immediately if one looks at the *history* of Tib and Tibbles. Therefore, the Location-Principle for the Identity of Material Objects cannot be right – at least not in the completely general way in which it has been formulated above.

One might, of course, draw a different moral from the story of Tib and Tibbles – that Tib is not a material object – which, if true, would render the Location-Principle, as formulated above, inapplicable to Tib and Tibbles; *or* that Tib does not exist at t – which, if true, would again make the Location-Principle inapplicable to Tib and Tibbles; *or*, since the example involving Tib, Tibbles, and time t is an entirely arbitrary example, that no "arbitrary undetached part" of any material object ever exists;[26] *or* that although Tib and Tibbles are at t existing material objects and do not have the same properties, they are nevertheless identical – as the Location-Principle requires, but contradicting the Leibniz-Principle; *or* that the identity-relation is not transitive – which can be argued for in the following (to my mind: confused) way: if t´ is a time before Tibbles lost its tail, then Tibbles at t´ is identical with Tibbles at t, and Tib at t is identical with Tib at t´, and Tibbles at t is identical with Tib at t (this is what the Location-Principle requires for the at t existing material objects Tib and Tibbles, isn't it?), but Tibbles at t´ just isn't identical with Tib at t´ (contradicting the transitivity of identity). None of these reactions, which seek to preserve the Location-Principle for the Identity of Material Objects, seems to me at all attractive. The best reaction to the Tib-and-Tibbles-objection, it seems to me, is to give up the Location-Principle.

The following principle, however, is immune against objections of the Tib-and-Tibbles type:

26 Cf. van Inwagen (1981).

The Fortified Location-Principle for the Identity of Material Objects
If O´ and O are material objects that both exist at some time and both exist at the same times, and if the place in space of O´ at every time of O´s existence is identical with the place in space of O at the same time, then O´ is identical with O.

But from the point of view of the present position on the places in space of material objects, there seems to be an objection of a different type even to the Fortified Location Principle, and even if one leaves modal considerations quite out of the picture. Let Tibbles-without-Proto be defined as Tibbles *without* a certain subatomic particle – the proton Proto – which, in fact, is in Tibbles at time t. Tibbles-without-Proto and Tibbles both exist at some time[27] and both exist at the same times. It also seems that the place in space of Tibbles-without-Proto at every time of this cat's existence is identical with the place in space of Tibbles at that time: there seems to be no reason to assign at any time of their simultaneous existence different places in space to Tibbles-without-Proto and Tibbles – considering that it is *we*, on the basis of our consciousnesses-*cum*-intentionality, who assign these places. But Tibbles-without-Proto is nevertheless non-identical with Tibbles – because their masses at t are non-identical, which, *according to the cloud-of-material-points hypothesis*, is equivalent to saying that their material-point-sets at t are non-identical. For Proto is at t in Tibbles, as we have supposed.

But this objection is, as a matter of fact, ineffectual. Take any moment t at which two material objects, O and O´, exist and which is such that VOL(t, O´) = VOL(t, O). On the basis of what has just been shown in the previous section, it follows that MAP(t, O´) = MAP(t, O). Therefore, if the material-point-set at t of Tibbles-without-Proto and the material-point-set at t of Tibbles are different – *and they are different* –, then their places in space at t must be different, too, and not identical as has been presumed in the above objection. Although it is indeed *we* who assign the places in space (qua localized volumes of space) to Tibbles-without-Proto and to Tibbles at t, we are bound by the strictures laid down in the Constitution-Statement itself to assign *different* places in space at t to these material objects.

27 Some will contend that the objection can be nipped in the bud by denying existence to Tibbles-without-Proto. But, to my mind, Tibbles-without-Proto exists at some time, since Tibbles exists at some time. For how could Tibbles ever exist without Tibbles-without-Proto existing at the same time?

6. The objection from physicalism

It remains to discuss some objections to the results presented in this paper. Perhaps the most important one of these objections is *the objection from physicalism*.

I have argued that every material object is at any time of its existence not purely physical, since its spatial unity at any time of its existence is not of a purely physical nature. I have argued, moreover, that it is *not* determined on purely physical grounds which objects exactly are the material objects that are constituted at a given time, that, in other words, the material objects that exist (are constituted) at a given time do not supervene on basic physical reality. I have asserted, and made plausible, I hope, that the material objects that are constituted at a given time are, to a considerable extent, products of human consciousness: of the intentionality of human consciousness.

But, one might object, it cannot be true that the material objects that exist at a given time do not supervene on basic physical reality and are to some extent products of human consciousness. It cannot be true, because human consciousness itself supervenes on basic physical reality. Suppose that the material objects constituted at time t are to some extent products of human consciousness. Then the only other factor in their production is, certainly, basic physical reality. But human consciousness itself, it is asserted, supervenes on basic physical reality. Hence, since basic physical reality determines human consciousness and is the only factor besides human consciousness in the production of the material objects constituted at t, basic physical reality *ultimately determines all by itself* which material objects are constituted at t. Therefore, according to this argument, the material objects constituted at t *do* supervene on basic physical reality even though they are to some extent products of human consciousness.

The central premise of this counter-argument is that human consciousness supervenes on basic physical reality. On this basis, one can argue that *it is not true* that the material objects existing (constituted) at a given time do not supervene on basic physical reality *and* are to some extent products of human consciousness. But one may just as well – *at least* just as well – take *as premise* the assertion that the material objects existing at a given time do not supervene on basic physical reality *and* are to some extent products of human consciousness, and argue on this basis, contrapositively, that human consciousness does not supervene on basic physical reality. This, in itself, is an important result for the philosophy of mind, which is dominated by materialism in a way that can hardly be considered rational.

7. Is this idealism?

It is not, of course. Neither the mind-independent existence of space nor the mind-independent existence of matter is being denied. It is merely maintained that certain, not unimportant features of physical reality are underdetermined by basic physical reality and are co-determined in their ontological constitution by the human mind. Without the contribution of the human mind *all of physical reality* would just amount to *basic physical reality*, and basic physical reality, as far as its specifically *material* aspect is concerned, is, according to the (by and large) classical picture that was employed in this paper, just the lawfully unified totality of material points in motion. The picture of basic physical reality which is induced by quantum physics is rather less clear than this; but that it is also true from the quantum-physical point of view that *material objects* – three-dimensional material space-takers, in particular, the macroscopic *things* (*living* and *non-living*) that are the immediate intentional objects of our conscious experience of the physical world – neither belong to basic physical reality nor are determined by it *seems undeniable.*

"If it's not full-fledged idealism," one might respond, "yet it surely is at least *partial idealism.* And partial idealism is already absurd enough. According to your position, the dinosaurs, existing long before human beings made their first appearance in natural history, must have been constituted, in part, by *us.* And are there not countless material objects which perfectly exist though they are nobody's intentional object? And what, precisely, is meant if it is said that material objects are constituted by *us*? By all of us? By the experts? By *you*? One answer is as absurd as the other." My responses, in turn, to these objections are the following:

(1) Truthfully postulated material objects – like the dinosaurs that once existed but do no longer exist, or material objects that exist though they are nobody's intentional object (it is meant: nobody's *evidentially given* intentional object) – have at the times of their truthfully postulated existence truthfully postulated places in space. We provide them with these places in space – they do not have them *an sich* –, but of course not in the same manner as a place in space is provided by us for an actually seen and/or felt – hence *to a high degree evidentially given* – individual material object. We provide them with places in space more or less *unspecifically*, corresponding to the manner in which those material objects are objects for us: in the manner of *not being evidentially given to a high degree*, perhaps in the manner of *not being evidentially given at all* (though still truthfully postulated).

(2) The – partial – constitution of material objects is not normally an activity we carry out *in person*; it is normally achieved automatically (except

for tricky cases[28]) by our – by each one's – *cognitive life* (that is, one's stream of consciousness, *plus* the dispositions that are inherent in or come to be attached to it). Hence it would, indeed, be a more proper way of speaking to say that a material object *is constituted* (impersonally) in our – in each one's – consciousness[29] than to say that *we* constitute it. Nevertheless, properly understood, the latter phrase is not objectionable, and in particular, understanding the "we" *truly collectively* is not objectionable. For with regard to our cognitive lives (in which the constituting comes about), we human beings are in some respects – namely, the respects relevant for constituting material objects – rather similar to each other, and it needs to be emphasized (against the accusation of idealism): our cognitive lives interact with a single mind-independent basic physical reality in which we all are similarly rooted; this is what makes our cognitive lives similar to each other in the respects relevant for the constitution of material objects. The result is a common world of material objects – this stone for you is the same stone for me (which is not to say that there is no room for disagreements). However, this common world of material objects is neither a part of basic physical reality nor determined by it. It is, as a matter of fact, not of a purely physical nature, but a partly mind-constituted *Lebenswelt* (to appropriate a famous Husserlian term).

8. Two easy ways out?

There seem to be two easy ways of escaping my argument that material objects are at no time of their existence purely physical because their spatial unity is at not time of their existence of a purely physical nature. What makes these ways seem easy is that they do not deny the Isolation-Assertion; they are reactions to the Isolation-Assertion – accepting its truth – but they are different from the way of dealing with this assertion that I have favored in this paper.

28 Such cases are material objects – for example, orchards – that even to the naked eye have a lot of empty, or at least *airy*, space "to them." How much of that space belongs to them at a given time – to their place in space at that time –, and how much of it does not? This cannot be decided automatically. (To the extent it is left undecided, the objects fail to be constituted, fail to exist; but, in practice, we often take near-existence as existence-near-enough.)

29 Note that Edmund Husserl, the great 20th-century idealist, often used the reflexive expression "konstituiert sich" for what is intended here by "is constituted." The German expression can be literally translated as "constitutes itself." But unfortunately the English equivalent emphasizes agent-reflexivity, which, however, is a meaning component not at all intended by Husserl (quite in accordance with German usage). Thus the impersonal passive expression "is constituted" is the proper rendering of "konstituiert sich."

I have proceeded on the assumption that some material objects exist at some time and that every material object has at every moment of its existence a *precisely defined* spatial unity (that is, a *spatial unity* in the usual, unqualified sense, which unity requires it to have at every moment of its existence a *precisely defined, determinate* place in space); that unity, however, (as I have argued) is never of a purely physical nature – and therefore: *no material object is at any time of its existence purely physical.*

The vagueness-reaction, on the contrary, maintains, vis-à-vis the Isolation-Assertion, that although some material objects exist at some time, no material object has at any moment of its existence a precisely defined spatial unity – but, nevertheless, *all material objects are at all times of their existence purely physical.* There is, according to the vagueness-reaction, no reason to give up this ontological tenet.

The elimination-reaction, in turn, maintains, vis-à-vis the Isolation-Assertion, that because no material object has at any (hypothetical) moment of its existence a precisely defined spatial unity, no material object (in fact) exists at any time – and therefore (trivially and vacuously): *all material objects are at all times of their existence purely physical.*

The vagueness-reaction and the elimination-reaction – they are in opposite ways *metaphysically radical* – have, as a matter of fact, nothing to recommend them – except, of course, that they avoid "idealism." For it seems undeniable that even some *macroscopic* (hence composite) material objects exist at some time (and not just living organisms, as van Inwagen believes,[30] but also artifacts – for example, the painting called "Mona Lisa" – and non-living natural objects – for example, the moon). It also seems undeniable that there is no such thing as an at a time t existing material object without a precisely defined spatial unity at t (or, in other words, with only a fuzzy spatial location at t). Lest this seem ontological dogmatism, here is an argument:

There will be, of course, an *appropriate* set of material points M[31] and a many-membered set Ω of localized volumes of space, which is such that each member V of it is such that all the material points in M are at t in V and such that V is otherwise empty at t. But the pair-set {M, Ω} does not constitute at t a material object with a fuzzy spatial location. For if it con-

30 Van Inwagen (1995) makes an exception for living organisms, but otherwise advocates eliminativism with regard to composite material objects – for reasons that seem to me, on the whole, closely related to the basic ontological assumption of the elimination-reaction: *lack of unity entails lack of existence.* This assumption is quite correct (but note that it is *negated* by the vagueness-reaction). What is incorrect is the other assumption of the elimination-reaction (*shared* by the vagueness-reaction): that all material objects lack unity at any moment at which they exist (though it is true that all material objects *would* lack unity at any moment at which they exist *if* their unity had to be built solely on physical grounds).

31 M is "appropriate" in the sense of being fit to *constitute a material object.*

stituted one, would not {M, Π} constitute at t *another* such object? – where Π is any *proper subset of* Ω with at least two members, *or* any *proper superset of* Ω which is such that each member of it is a localized volume of space V′ such that all the material points in M are at t in V′ and such that V′ is otherwise empty at t. In other words, once one gets started with spatially vague material objects, there are just too many spatially vague material objects around: infinitely many, all constituted (hence existing) at the same time t, all based on the same set of material points M, some of them spatially vague at t to a high degree (if Π is very much diversified), others to a low degree (if Π is not much diversified). Instead of living with this totally unnecessary infinity of vague material objects, it is certainly better to accept that there is just *no* spatially vague material object *at all* constituted (i.e., existing) at t on the basis of M.

But might one not reduce the flood of infinitely many spatially vague material objects that are prima facie constituted at t on the basis of M *to just one* by selecting one of them, *the true one* (so to speak), discarding all the others? But *who* would do the selecting? *We*, of course, and if it is not to be a totally arbitrary choice that we are making, then our consciousness-*cum*-intentionality, hence our experience of material objects, would have a large say in it. In other words, we are back to "idealism" – the very thing that the vagueness-reaction set out to avoid.

References

Lewis, D. (1999) 'Many, but almost one' in: Lewis, D. *Papers in Metaphysics and Epistemology*, 164-182, Cambridge: Cambridge University Press.

Unger, P. (1980) 'The Problem of the Many', *Midwest Studies in Philosophy* 5, 411-467.

Van Inwagen, P. (1981) 'The Doctrine of Arbitrary Undetached Parts', *Pacific Philosophical Quarterly* 62, 123-137.

Van Inwagen, P. (1995) *Material Beings*, paperback edition, Ithaca and London: Cornell University Press.

SERIOUS ENDURANTISM AND THE STRONG UNITY OF HUMAN PERSONS

E. JONATHAN LOWE

Abstract

In this paper, I distinguish between a rather weak form of endurantism which is consistent with a so-called *B-theoretic* approach to the philosophy of time and a much stronger form — one that I call *serious* endurantism — which is committed to a kind of 'presentism'. It is this latter form of endurantism that I favour. According to this view of time and persistence, the unity of time itself depends upon the successive but 'overlapping' existence of simple substances, each possessing primitive or irreducible identity over time — rather as the unity of a rope depends upon the existence of the individual overlapping fibres that make it up, even though no single fibre may run throughout the whole length of the rope. The passage of time, on this view, consists in the *absolute* coming into existence and passing away of particular entities, whether these be individual substances themselves or their 'modes' — that is, their individual properties and relations — and time has no reality apart from such absolute existential change. Thus, the passage and unity of time both ultimately depend ontologically upon the existence of simple, changeable substances — substances which, in virtue of their simplicity or non-compositeness, have a particularly 'strong' form of unity. Furthermore, philosophical reflection and empirical scientific information together suggest that the only known or likely candidates for the status of 'simple substances' in this sense are conscious beings such as ourselves, that is, *persons*. The paper concludes with some brief speculations of a theological character.

1. The thesis of 3D/4D equivalence

In a number of recent papers, Storrs McCall and I have defended what we call the *thesis of 3D/4D equivalence* with regard to theories of persistence and change (see McCall and Lowe 2003 and McCall and Lowe 2006; also Lowe 2005). According to this thesis, 'four-dimensionalist' and 'three-dimensionalist' accounts of persistence and change do not present metaphysical pictures of temporal reality that are fundamentally different, contrary to the suppositions of many of their respective advocates. The four-dimensionalist says, typically, that objects persist through time in virtue of possessing different temporal parts at different times, and that they undergo change over time in virtue of the different properties and relations borne by those parts at different times. For example, a banana that changes from being green to being yellow does so in virtue of possessing first a green temporal part and later a yellow one. Qualitative *change* is thus conceived as consisting in qualitative *difference* between a changing object's successive temporal parts. I say that this is what four-dimensionalists 'typically' say, because not all of them subscribe to this 'perdurantist' view of persistence, whereby an object such as a banana is a four-dimensional 'spacetime worm', composed of successive temporal parts. For there is also the 'stage-theoretic' version of four-dimensionalism, according to which, although such spacetime worms exist, an object such as a banana is in fact a momentary temporal part or stage of such a worm, and 'persists' over time only in the sense that it has earlier and later temporal 'counterparts', in the form of other temporal parts or stages of the same spacetime worm (see, notably, Sider 2001 and Hawley 2001). However, it seems clear that these two versions of four-dimensionalism do not present metaphysical pictures of temporal reality that are fundamentally different, since they are both committed to an ontology of spacetime worms and their momentary temporal parts or stages, and differ only over the question of whether to identify 'ordinary objects', such as bananas, with spacetime worms or with their momentary stages.

But *three*-dimensionalism, on the other hand, certainly seems, at first sight, to present a radically different metaphysical picture of temporal reality. According to this view of persistence and change, persisting objects, such as bananas, are certainly *not* composed of momentary temporal parts or stages, nor are they identical with the latter: rather, they are, as the somewhat obscure phrase has it, 'wholly present' at every time at which they exist, and undergo qualitative change by possessing, in a perfectly non-derivative and direct sense, different properties at different times. On this view, then, our banana is *itself* green at one time and yellow at a later time, rather than possessing first a green and later a yellow temporal part, or being replaced at the later time by a yellow counterpart of itself. This view, of course, is supposed to be vulnerable to the problem of 'temporary intrinsics', because it is alleged that it is committed to regarding supposedly intrinsic properties, such as colour-properties, as being, in reality, *relational* ones, consisting in peculiar relations to times, such as the *green-at* relation and the *yellow-at* relation (see, especially, Lewis 1986, 202–204). However, there are many responses available to the three-dimensionalist to counter this complaint, which certainly doesn't appear to be a decisive one, so I shall say no more about it here (see further Lowe 2002, ch. 3). More to the point, for current purposes, is the fact that three-dimensionalism, as just described, certainly *seems* to present a radically different view of persistence and change from that presented by either version of four-dimensionalism.

However, that such a radical difference exists is precisely what is denied by the thesis of 3D/4D equivalence. This thesis does not maintain that three-dimensionalism and four-dimensionalism are not *different* theories of persistence and change — for they manifestly are. It maintains, rather, that the difference between them is not metaphysically fundamental, because the two approaches are, in an important sense, *equivalent.* They are equivalent in the sense that a *one-to-one correspondence* can be set up between the entities invoked by the two different approaches and a truth-preserving *translation-scheme* between their respective statements concerning the persistence and change of objects over time. Where they differ is only with regard to the 'order of construction' of various *non-basic* elements of the respective approaches, such as 'ordinary' macroscopic persisting objects — things like bananas. Thus, where the four-dimensionalist of the perdurantist variety regards such an object as being *composed* of momentary temporal parts, each of these parts being composed of the simultaneous momentary temporal parts of certain fundamental particles, the three-dimensionalist regards such an object as being *constituted*, at any given time, by a set of such particles and thus as *coinciding*, at that time, with the mereological sum or fusion of those particles.

Call the object in question — an object such as our banana — *O*, and let *S* be the set of fundamental particles which, according to the three-dimensionalist, constitute *O* at any given time *t*. Finally, let '<*S*, *t*>' denote what the perdurantist four-dimensionalist would call *O*'s momentary temporal part at time *t*, namely, the mereological sum or fusion of all the momentary temporal parts at *t* of the particles belonging to *S*. Then, where such a four-dimensionalist says that *O* exists at *t* in virtue of having <*S*, *t*> as a temporal part, the three-dimensionalist says that *O* exists at *t* in virtue of being constituted by the members of *S* at *t*. Both theorists acknowledge the existence of *O*, *S*, and *t*, but differ over how *O* is to be 'constructed' from *S* and *t*. Most importantly, for every object included in the four-dimensionalist's ontology, the three-dimensionalist can recognize a corresponding one. Thus, corresponding to the four-dimensionalist's momentary temporal part of *O* at *t*, <*S*, *t*>, the three-dimensionalist can acknowledge the existence of an instantaneous three-dimensional object, *O**, which exists only at *t* and exactly coincides with *O* at *t*, being constituted at *t* by the members of the same set of fundamental particles, *S*, whose members constitute *O* at *t*. According to the three-dimensionalist, *O* and *O** are *coinciding three-dimensional objects*, whereas according to the perdurantist four-dimensionalist, the object 'corresponding' to *O**, namely, <*S*, *t*>, is the *momentary temporal part* of *O* at *t*. However, there is *no real difference* between *O** and <*S*, *t*>, just a difference in the theoretical roles they play in the two different approaches.

2. Two forms of endurantism

Now, the key claim of three-dimensionalism — otherwise commonly known as *endurantism* — is that objects do not persist through time in virtue of possessing successive temporal parts, for on this view persisting objects simply do not *have* temporal parts at all. However, as we have just seen, three-dimensionalism can maintain this, perfectly consistently, while also admitting the existence of instantaneous three-dimensional objects which are *not really any different* from the four-dimensionalist's momentary temporal parts — it's just that, for the three-dimensionalist, such an object *coincides*, at an instant, with a persisting object, both being *constituted* at that time by the same set of fundamental particles, whereas, for the four-dimensionalist, such an object is one of the momentary temporal parts of which the persisting object is *composed* (on the perdurantist view) or has as a temporal 'counterpart' (on the stage-theoretic view). Such a three-dimensionalist may regard a persisting object as being *extended over time*, just as a four-dimensionalist may — at least according to the perdurantist version of the latter approach — but must simply deny, as is perfectly feasible, that an object can be extended over time only in virtue of possessing temporal parts of shorter extents that are extended over sub-intervals of the time-interval over which that object is extended.

This, certainly, is *one* form of endurantism, albeit one that — or so I claim — is not *fundamentally* different from perdurantism, for the reasons explained earlier. However, there is also another and more radical form of endurantism — one that I propose to call *serious* endurantism — which not only denies that objects persist through time in virtue of possessing successive temporal parts, but also denies that they are in any sense *ex-*

tended over time. According to this view, time is simply not a *dimension* of reality in which things can be extended, in the way in which each of the three dimensions of space are. Yet, both versions of *four*-dimensionalism — both perdurantism and the stage theory — are committed to regarding time in precisely this dimensional way, as indeed their collective name suggests. This, at root, is because both versions are committed to the existence of objects — so-called 'spacetime worms' — which are *composed* of momentary objects existing at different times. (As we have seen, the two versions differ only over the question of whether or not 'ordinary objects', such as bananas, are to be identified with such spacetime worms themselves or rather with their momentary temporal parts.)

The key point here is that a composite entity can be *composed* of certain other entities only if those entities stand in certain *real relations* to one another. By a 'real relation' here I mean a so-called *external* relation, such a relation being one that does not obtain between its relata purely in virtue of their intrinsic properties. The contrast here is with an *internal* relation, which *does* obtain for this sort of reason. A paradigm example of an internal relation would be the *same height* relation, which obtains between its relata purely in virtue of their respective heights. Spatial relations are, according to most metaphysicians, paradigm examples of external relations, since a spatial relation between two objects, such as the distance between them, can be varied independently of their intrinsic properties. Many metaphysicians also regard *causal* relations as external. Now, a real relation can obtain between two objects only if both objects *exist* and, more importantly, *co*-exist. By 'co-exist' I do not specifically mean *exist at the same time*, but only *exist together*, in a more general sense, of which the notions of existence at the same time or in the same place are just restricted versions. However, the crucial difference between 'serious' endurantism, as I understand it, and any version of four-dimensionalism is that only the latter regards all *temporal* relations, like spatial relations, as being external relations. The serious endurantist denies that *cross*-temporal relations are external precisely because he or she does not believe that objects whose existence is confined to different times *co-exist*, in the most general sense of this expression. For such a theorist holds that an object's 'coming into' and 'going out' of existence are to be understood literally as involving an *absolute* existence-change: after such a change, the object in question simply *does not exist at all* — does not belong to the sum total of reality. We see, then, that such a theorist cannot by any means countenance the existence at *any* time of an object composed of momentary parts whose existence is confined to *different* times, because, for such a theorist, those supposed parts do not in any sense *co-exist*, and so cannot be the relata of real relations. Only a theorist who believes that cross-temporal relations are real relations and thus that momentary objects existing at different times do genuinely *co-exist* can maintain that there are temporally extended objects *composed* of such momentary objects, in the way that the four-dimensionalist's spacetime worms are conceived to be. For, as I said earlier, *composition* requires a *real relation* of some kind between the component parts of a composite object — whether the relation in question be spatial, causal, or indeed temporal.

The upshot of all this is that the thesis of 3D/4D equivalence applies only to the less radical form of endurantism identified earlier — not to *serious* endurantism. And it is the latter theory that I myself favour. I do so for a number of interrelated reasons. One is that I do not believe that there are real cross-temporal relations — that is to say, I do not believe that 'separation in time' is anything like 'separation in space', or *distance.* The notion that the past is like a distant country, while a strangely seductive one, is, I think, fundamentally misconceived. At bottom, this is because I take the notion of 'passing out of existence' quite literally, as involving an *absolute* ceasing to be, rather than merely the occupancy of another and earlier temporal 'location'. The four-dimensionalist certainly doesn't take this notion seriously: for such a theorist, Julius Caesar is no less *real* now than he was when alive — he is just confined to a region of spatiotemporal reality that is far removed from our own. Of course, this may seem to place me in conflict with the tenets of the special and general theories of relativity — and perhaps indeed I am, at least as those theories are customarily interpreted. However, there is plenty of leeway for reinterpreting the empirical data commonly advanced in support of those theories in ways which are consistent with the metaphysics of time that I am now advocating (see, for instance, Tooley 1997, ch. 11). We must not confuse the metaphysical assumptions of scientific theorists with the empirically confirmable contents of their theories.

3. Temporal passage and presentism versus eternalism

I now need to say a little more about my conception of *time.* In effect, I am rejecting what is commonly called a 'B-theoretic' or 'eternalist' conception of time, which regards different times as being 'equally real' locations along a dimension of reality that is akin, ontologically, to the three dimensions of space. Indeed, I do not wish to *reify* time or times at all. I accept *neither* an 'absolutist' *nor* a 'relationalist' view of time. I reject the former because it reifies time itself as something that supposedly exists independently of individual, changeable things. But I also reject the latter, because it regards time as a system of *real relations* between things and events and, as I have just explained, I do not consider that there are real cross-temporal relations, analogous to the real distance relations between things in space. Talk about time and times, as I see it, is just an abstraction from talk about *things changing*. Of course, it is commonly said there are several different varieties of change, such as substantial, qualitative, and relational change. However, in my view, the crucial kind of change that underpins all talk of time and its passage is *existence-change* — coming into or passing out of existence, in an absolute sense. In fact, I consider that all of the varieties of change just mentioned are species of existence-change, differing only in respect of the type of entity whose existence is concerned. Thus, substantial change is the existence-change of *substances* — that is, persisting objects — while qualitative change and relational change are, respectively, the existence-change of individual *qualities* and the existence-change of individual (external) *relations*. By an 'individual' quality or relation in this context I mean what are commonly called monadic and relational *tropes* or *modes*. Thus, I understand the qualitative change of our banana from green to yellow as consisting the passing out of existence of its green trope or mode and the coming into existence of its yellow one. As I see it, all talk about time's 'passage' is just a compendious way of talking generally about the coming into and passing out of existence of entities of various different types. Consequently, if it were not for such existence-changes, all talk of time and its passage would be empty. It makes no sense to suppose that there might be a 'period of time' during which no such existence-change occurred.[1] To suppose otherwise is to turn a mere abstraction into something robustly real in its own right. However, provided that we do understand this, there is no harm and much convenience in talking about time and times, as I have done hitherto and will continue to do.

It may be asked: given that I reject a B-theoretic view of time, do I therefore accept an A-theoretic one — and, more specifically, do I adopt a so-called *presentist* view of time? Well, much depends here on what one takes the latter view to involve. Some self-styled presentists describe their position as being one according to which *the only real moment of time is the present moment*. But I certainly can't accept this, since I hold that all talk about 'moments of time' involves a mistaken reification of mere abstractions. However, there are presentists who would seemingly agree with me that talk about 'times' is talk about abstractions, but mean thereby that times are *real but abstract objects*, such as maximal consistent sets of present-tensed propositions, constituting different 'representations' of present reality — by analogy with so-called 'ersatzist' accounts of the ontological

1 I disagree, then, with Sydney Shoemaker's main contention in Shoemaker 1969. For discussion, see Lowe 2002, 247–9.

status of 'possible worlds', according to which these are maximal consistent sets of propositions constituting different representations of *actual* reality (see Lowe 2002, 42–3). Now, since I certainly believe in the existence of propositions and regard these as abstract entities which exist timelessly, I can certainly countenance the existence of such maximal consistent sets of present-tensed propositions. I am not at all convinced, however, that we need to refer to or quantify over such abstract entities, identifying them with 'times', in order to make sense of our temporal language.

Against this, it is sometimes argued, for instance, that without such quantification, even given a rejection of the B-theoretic approach, we can make no sense of the present truth of a past-tense statement such as 'There were two queens of England named "Mary"' — because we cannot explain the present truth of this statement simply by contending that the *present-tense* proposition 'There are two queens of England named "Mary"' *was* true.[2] For the latter wasn't *ever* true, since the two queens in question lived many years apart and never *at the same time*. Consequently, it is alleged, we need *somehow* to quantify over 'times', even if only conceived in some 'ersatzist' fashion, in order to resolve this difficulty.

However, I believe that this apparent problem has a relatively simple solution, provided that we don't subscribe to the pervasive dogma that the so-called 'existential' quantifier, '∃', expresses *existence*. In my view, this is more aptly called the 'particular' quantifier and existence can be perspicuously expressed only by means of a primitive or indefinable *first-level predicate*, 'exists' (see further Lowe 2003 and compare McGinn 2000, ch. 2). Then we can say that the past-tense statement 'Two queens of England named "Mary" *existed*' may be formally regimented as follows:

(1) $\exists x \exists y$(x existed and was a queen of England named 'Mary' and y existed and was a queen of England named 'Mary' and $x \neq y$).

It is important to note, at this juncture, that statements of *identity* and their negations are essentially tenseless or timeless — the underlying reason for this being that identity is not an external relation. (It is true that philosophers commonly distinguish between 'synchronic' and 'diachronic' identity, but it is a mistake to think that what is at issue in such talk is two different kinds of *identity*, each involving a different kind of temporal relation between its relata: for identity is just a *formal* relation which necessarily obtains between any thing and itself, irrespective of when or where that thing may exist.) Now, I take it that a past-tense statement of the form '*a*

2 I owe this example to Craig Bourne.

existed and was a queen of England named "Mary"' is *presently* true just in case the present-tense proposition '*a* exists and is a queen of England named "Mary"' *was* true — a principle that can be used to construe each of the first two conjuncts in the doubly quantified statement (1) cited above. The basic point here is that when *counting* things we do not have to presume their *existence*, much less their *co*-existence — after all, did not the Red Queen in *Alice Through the Looking-Glass* claim to think of six impossible things before breakfast? The common misconstrual of the particular quantifier, '∃', as expressing existence obscures this important fact and is responsible for the pseudo-problem that we have just been dealing with. Statements of number can indeed be expressed using this quantifier and the identity sign in the standard fashion, but *without existential import.*

Returning to the question of whether my view of time is a 'presentist' one, I would say that it is so in the following sense: if something *did* exist but does not *presently* exist, then it has *absolutely* ceased to exist and so does not belong any more to the sum total of reality. According to my view, the very *content of reality* continually changes over time, which is why we can talk as we do of time 'passing'. By contrast, on an 'eternalist' or 'B-theoretic' conception of time, to say that something *did* exist but does not *presently* exist is only to *locate* its existence at an earlier time, much as one may locate something's existence at another *place.* On this view, existence as such is essentially timeless: not so on my view, at least as far as concrete, changeable things are concerned.

4. The problem of the unity of time

Now, however, I need to address an important but neglected issue in the philosophy of time — namely, its *unity*. The question at issue may be put in this somewhat grandiloquent way: in virtue of what is the world *one* world in time? What makes it the case that all of the times in 'our' time-series do genuinely belong to *one and the same* time-series, as opposed to different and disconnected ones? Of course, to express the question in this way is to use the reificatory language of 'time' and 'times' — but, as I mentioned earlier, this is harmless enough and convenient, provided that we do not take it too seriously. Our question does not really arise for those who hold an *absolute* conception of time, for whom the time-series is just an essential feature of time itself. Nor does it really arise for those who hold a *relational* conception of time, because for them the time-series emerges automatically from the set of cross-temporal relations between things and events out which time, according to them, is constructed. And yet it does seem to be a deep and interesting question — and this in itself suggests that both absolutists and relationalists, to the extent that the question doesn't really arise for them, have not penetrated to the bottom of the nature of time.

In order to motivate an answer to our question, I propose the following simple thought-experiment. What would happen if *everything* presently existing were to go out of existence simultaneously? I think that the correct reply is that this would be *the end of the world*, and therewith *the end of time*. It makes no sense, I think, to say that there could be things existing *after* such a total existence-change. Similarly, it makes no sense to say that certain things *did* exist *prior* to a total existence-change. If there was a time at which everything then existing *did not exist prior to that time*, then that was, *ipso facto*, the *first* moment of time. However, I have already said that the passage of time requires the continual coming into and passing out of existence of at least some things, whether these be substances or their qualities and (external) relations. Putting these two thoughts together, we can conclude that, while the *passage* of time requires something to come into or go out of existence at each moment of time, the *unity* of time requires that, for any two adjacent periods of time, there be at least something that persists across the boundary between those periods. More specifically, we can say that, for any such boundary, a *substance* must persist across it, because individual qualities and relations cannot persist unless their bearers do — for they are dependent for their *identity* upon their bearers and consequently cannot be transferred from one bearer or set of bearers to another.[3] Even more specifically, we can say that a *simple* or *non-composite* substance must persist across such a boundary, because a *composite* substance (one that has other substances as its component parts) cannot persist unless at least some of its parts do. In short, then, the unity of time rests on the persistence of simple substances in an 'overlapping' fashion, like the fibres in a rope. Although no single fibre need extend throughout the whole length of a rope, there can be no cross-section of it through which at least *some* fibres do not extend, on pain of dividing the one rope into two separate ropes. Similarly, then, although no single simple substance need persist for all time in order to sustain the unity of time, there can be no moment of time through which *no* simple substance persists.

This, if correct, is a remarkable conclusion. Of course, it might be challenged. The most obvious way to try to challenge it is to argue that there are and must be some real external relations which obtain over time. If there were real cross-temporal relations analogous to spatial relations like distance, then they would do the job. But I have contended that there

3 I defend this view concerning the identity conditions of individual qualities and relations in, *inter alia*, Lowe 2006, Part II.

are none. However, I mentioned earlier that many philosophers think that *causal* relations are real external relations: and most of them also hold that such relations obtain between entities — individual events — *existing at different times*. *Causal* theories of time assume precisely this. Naturally, I cannot accept the reality of causal relations conceived in this fashion: for if the cause of an event, *x*, is supposed to be another event, *y*, which has already *ceased to exist* by the time that *x* exists, then *x* and *y* are not, according to my view, parts of *the same sum total of reality*, so that no real external relation can obtain between them. In point of fact, I think that all talk of 'event-causation' is merely a convenient *façon de parler* and that individual *substances* are the only entities that really *cause* anything.[4] That is to say, all causation is fundamentally *substance-causation*, which is a matter of individual substances exercising or manifesting their various causal powers and liabilities. Consequently, no appeal to facts about causation can undermine my earlier conclusion that it is the *persistence of substances* that ultimately underpins the unity of time. For how could a substance, by exercising its causal power at an earlier time, have any effect on another substance at a later time, unless at least *some* substance persisted between those times to *transmit* this causal influence? A *complete* existence-change between those times would terminate any such transmission process and thereby prevent the causation of the putative effect. Causation, I believe, works something like this: when a substance exercises one of its causal powers, its *immediate* effect is simultaneous with that exercise and consists in an existence-change in that or one or more other substances — either a substantial change, or else and more commonly a qualitative or relational one. The affected substances may then persist in their altered state for a while until they in turn exercise one or more of their causal powers. This is how causal influence is propagated across time. Strictly speaking, all of the propagation *across time* is secured by the persistence of substances. Hence, once more, no appeal to causal considerations can threaten my claim that it is ultimately the persistence of substances that is responsible for the unity of time.

Remember, however, that my claim is, more specifically, that it is the persistence of *simple* substances that underpins the unity of time. I see no way to weaken this claim so as to allow merely composite substances to shoulder this burden, since they can persist only provided that at least some of their *substantial parts* do — and an infinite downward regress of substantial composition does not look to be either metaphysically feasible or empirically plausible. This leaves us, though, with another pressing

4 I defend this view of causation at length in Lowe 2008.

question: what *are* the simple substances that underpin the unity of time? It is to this question that I now finally turn.

5. The strong unity of human persons

My reflections in this last section of the paper will be somewhat speculative, I have to confess. What I want to suggest, perhaps rather surprisingly, is that the most likely sort of candidates that we know for the status of simple substances are none other than *beings like ourselves* — that is to say, *persons* or, speaking a little more generally, *subjects of experience.* It might be thought that more plausible candidates would be the so-called *fundamental particles* of modern physics — things such as quarks and electrons. However, it is rather questionable whether these are really properly thought of as being *persisting objects.* According to quantum field theory, they are more like disturbances in widely pervading energetic fields, so that, for example, talk of a 'single' electron 'moving through space' is really just a picturesque way of describing the manner in which an energetic field pervading a region of space undergoes a regular succession of modifications in its field density along a certain spatial path. Perhaps it will be now be objected that if that is so, then, after all, the account of causation offered earlier, appealing to the transmission of causal influence by persisting substances, is incompatible with modern physics. That *may* be so: but if it is, and if the picture presented by the field-theoretic approach is correct, then perhaps the conclusion should be that causation is, as Bertrand Russell famously held, a pre-scientific notion for which modern physics has no real use. However, the more fundamental difficulty that we are faced with as metaphysicians, in the light of developments in modern quantum physics, is that *no one really knows* how quantum theory should be interpreted as a putative account of the ultimate nature of physical reality. The problem is not a *lack* of interpretations, but rather a *plethora* of them, all mutually incompatible and many really quite bizarre. I propose, therefore, not to appeal to or rely on *any* of these interpretations. In my view, confusion and disagreement about the proper interpretation of quantum theory does nothing to undermine the notion of a persisting substance as such, which seems to have a much securer foothold in our thought about reality than any passing scientific theory about the ultimate constituents of the physical world. Indeed, it is worth mentioning, in this context, that even the field-theoretic approach alluded to above does not really entirely abandon the notion of persisting substance — for, in effect, the *fields* themselves take on this status, since they are supposed to persist through time and undergo qualitative modification.

If one claim in metaphysics seems relatively unassailable, even in the light of contemporary physical science, it is that *we* exist and persist through time. But, I suggest, when we reflect upon our nature as persisting things, we find quite compelling reasons to believe that we ourselves qualify, if anything does, as *simple substances*. We are 'substances' (to use that old-fashioned word) inasmuch as we are persisting bearers of qualities and relations, in respect of which we continually undergo qualitative and relational change — for instance, whenever one of our thoughts or experiences gives way to another. But that we are *simple* rather than *composite* substances will appear much more controversial, or even just plain absurd, on first reflection. This is because it is commonly assumed that we are identical with our living organic bodies or, if not with our bodies as a whole, then at least with certain distinguished organic parts of them, such as — most obviously — our brains. Any suggestion that we are *simple* substances is likely to be associated with the supposedly untenable Cartesian conception of the self or ego as an immaterial and unextended — and therefore indivisible — substance. However, we have already seen that the notion of an object that is *extended* along some dimension and yet does not possess parts of lesser extent in that dimension is perfectly coherent: for this is precisely what an adherent of the 'weaker' form of endurantism discussed previously may say concerning an object's persistence through *time* — namely, that an object may be *extended* in time and yet lack temporal *parts*. Similarly, then, an object may be extended in *space* and yet lack *spatial* parts, that is, fail to be composed of smaller objects which occupy sub-regions of the region of space over which it is extended. Indeed, if an electron is properly thought of as a persisting object at all, then it would appear to extend over space in precisely this fashion: for, being a *fundamental* particle, it is not (according to current physical theory) composed of other particles, but nor does it occupy a mere *point* of space at any given time (since that would give it an infinite energy-density).

Elsewhere, I have developed a theory of the self according to which human persons, although not lacking physical properties altogether, are simple substances that are *not identical with* their organic bodies or any part of them (see, especially, Lowe 1996, ch. 2 and Lowe 2001). Call this *the non-identity thesis*. On this view, a person *inherits* certain physical properties from his or her body — such as its size, shape, mass, spatial location, and velocity — but is not *composed* of any of that body's parts. The non-identity thesis is supported by various considerations which indicate that the *identity-conditions* of persons and their bodies (and bodily parts) are quite different. But, above all, it is supported by what I call *the unity argument*. Very briefly, what this argument purports to show is that a person's conscious states — his or her conscious thoughts and experiences — all depend ontologically upon *that person* in a fashion in which they do *not* depend upon that person's body as a whole or on any distinguished part of it (such as the brain) as a whole. For example, all of *my* conscious thoughts and experiences depend ontologically upon *me*: *none* of them could exist if *I* did not exist. But it is not true of my body *as a whole*, nor of any part of it *as a whole*, that none of my conscious thoughts and experiences could exist if *it* did not exist. Consider, for instance, *my brain*. Although it is quite probably true that if I did not have *any brain at all*, none of my conscious thoughts and experiences could exist, it plainly isn't the case that all of my conscious thoughts and experiences depend upon my brain *as a whole* existing: for I could certainly have many of those thoughts and experiences while losing some relatively small part of my brain. Consequently, *I* cannot be identified with my brain *as a whole*. It may even be true that *each* of my conscious thoughts and experiences depends upon *some* part of my brain existing: but it would obviously be fallacious to infer from this that there is *some* part of my brain upon whose existence *all* of my conscious thoughts and experiences depend, in the way in which they clearly do all depend upon *my* existence. (To draw such an inference would be to indulge in a so-called 'quantifier-shift' fallacy.) The clear fact seems to be that the way in which a person's conscious thoughts and experiences depend upon that person's brain — namely, in such a fashion that *different* thoughts and experiences depend upon *different* parts of that brain, rather than *all* of them on the brain (or some distinguished part of it) as a whole — is quite different from the way in which they all depend upon *that person*. If all that linked the various different conscious thoughts and experiences of a single person were the various different dependencies that they have on different parts of the same brain, then those thoughts and experiences would not be linked *to each other* in the strongly unified way that they seem to be. Only their common dependency upon *the person* whose thoughts and experiences they are seems capable of sustaining this unity. Now, it is precisely the *composite* character of the brain and the body as a whole that ill-equips them for this unifying role, whence it is natural to conclude that a person or self can fulfil that role only because it is *simple* or *non-composite* in character.

There is, naturally, much more to be said about this line of argument for the simplicity of the self. Nor is it the only line of argument that supports that view of the self. Suffice it to say, for present purposes, that there are cogent, even if not utterly incontrovertible, grounds for thinking that persons or selves — and, more generally, subjects of experience — are simple substances, if they exist at all. There have, of course, been philosophers who have *denied* the existence of the self altogether. But in thereby denying their own existence, they would seem to be ruling themselves out of the present debate. It seems not unfair to ignore their existence, since they deny it themselves. All that I want to conclude for the time being, however tentatively, is that persons or selves seem to be the most likely candidates we know of for the status of simple substances.

Suppose, then, that persons or selves *are* the only simple substances. What would follow? Well, I argued earlier that the passage and unity of time depend upon the existence of simple substances. The obvious implication would be that they depend upon the existence of beings like ourselves — *minded* beings. Would that conclusion, if true, commit us to an *idealistic* view of time? I don't think so. The implication would not be that time has no reality outside the mind, only that it would depend upon the real existence of minded beings like ourselves. Even so, would we not have to conclude that there was no time before minded beings like ourselves came into existence — and does not this seem incompatible with the views of modern science regarding the distant past of the physical universe when, it appears, no conscious beings like ourselves existed? This, certainly, is an awkward question. But there is, of course, one obvious way to reconcile these scientific views with the conclusion that we are now contemplating: namely, to posit the existence of a minded Being who has existed since the very beginning of time itself. Perhaps, after all, there is *one* individual fibre that *does* run throughout the whole rope of time. However, if that everlasting Being is, as traditional theology maintains, an *unchanging* substance, then the implication will be that while the *unity* of time depends upon this Being, the *passage* of time depends upon the mutable nature of less permanent substances, whether these be conscious or unconscious ones.

References

Hawley, Katherine (2001) *How Things Persist*, Oxford: Clarendon Press.

Lewis, David (1986) *On the Plurality of Worlds*, Oxford: Blackwell.

Lowe, E. J. (1996) *Subjects of Experience*, Cambridge: Cambridge University Press.

Lowe, E. J. (2001) 'Identity, Composition, and the Simplicity of the Self' in: Corcoran, Kevin J. (ed.), *Soul, Body, and Survival: Essays on the Metaphysics of Human Persons*, Ithaca, NY: Cornell University Press, 139–158.

Lowe, E. J. (2002) *A Survey of Metaphysics*, Oxford: Oxford University Press.

Lowe, E. J. (2003) 'Identity, Individuality, and Unity', *Philosophy* 78, 321–336.

Lowe, E. J. (2005) 'Vagueness and Endurance', *Analysis* 65, 104–112.

Lowe, E. J. (2006) *The Four-Category Ontology: A Metaphysical Foundation for Natural Science*, Oxford: Clarendon Press.

Lowe, E. J. (2008) *Personal Agency: The Metaphysics of Mind and Action*, Oxford: Oxford University Press.

McCall, S. and Lowe, E. J. (2003) '3D/4D Equivalence, the Twins Paradox and Absolute Time', *Analysis* 63, 114–123.

McCall, S. and Lowe, E. J. (2006) 'The 3D/4D Controversy: A Storm in a Teacup', *Noûs* 40, 570–578.

McGinn, Colin (2000) *Logical Properties: Identity, Existence, Predication, Necessity, Truth*, Oxford: Clarendon Press.

Shoemaker, Sydney (1969) 'Time without Change', *Journal of Philosophy* 66, 363–381.

Sider, Theodore (2001) *Four-Dimensionalism: An Ontology of Persistence and Time*, Oxford: Clarendon Press.

Tooley, Michael (1997) *Time, Tense, and Causation*, Oxford: Clarendon Press.

WHY PARTHOOD MIGHT BE A FOUR-PLACE RELATION, AND HOW IT BEHAVES IF IT IS [1]

CODY GILMORE

1. Introduction

Compositional monism is the view that there is exactly one fundamental parthood relation, i.e., exactly one parthood relation that does not have an analysis in terms of some more natural[2] parthood relation. *Compositional pluralism* is the view that there are multiple such relations, perhaps associated with different ontological categories and possessing different, though overlapping, sets of formal properties.[3] (These terms are from McDaniel (2004 and forthcoming).)

One assumption is widely shared by philosophers on both sides of this dispute – viz., that there is just one fundamental parthood relation that can hold between one material object and another. I will make this assumption too, and I will call the relation in question *parthood$_m$*.

One of the central questions about this relation is *The Adicity Question*: What is the adicity of parthood$_m$? The most widely accepted answer is that parthood$_m$ is a two-place relation, with one slot for a part, another slot for a whole, and no further slots. Theodore Sider has defended this view and given it a name:

> Absoluteness: Parthood [or at least parthood$_m$] is a two-place relation; it does not hold relative to times, places, sortals, or anything else (2007: 70).

---

1 I am grateful to Yuri Balashov, Ben Caplan, Greg Damico, Scott Dixon, Maureen Donnelly, Hud Hudson, Kris McDaniel, and Adam Sennet for very helpful comments.

2 The notion of relative naturalness is from Lewis (1986a: 59-69): the more natural properties and relations are the ones that do a better job of 'carving nature at its joints' than the less natural ones; sharing them does more to make for genuine resemblance than the sharing of less natural properties and relations.

3 E.g., there may be one fundamental parthood relation that holds between states of affairs and their constituents, another that holds between complex universals and their constituents, another that holds between events and their constituents, etc. McDaniel (2004) defends compositional pluralism. Sider (2007) defends compositional monism.

The main rival to Absoluteness is

Three-Place Parthood (3P): Parthood$_m$ is a three-place relation that can be expressed by the predicate 'x is a part$_m$ of y at z'.

3P is popular with those who hold that at least some material objects are *multi-located in spacetime*, exactly occupying[4] (or being 'wholly present at') each of two or more spacetime regions.[5] These philosophers typically want to allow that a material object can have different parts$_m$ at different regions. (Or at different *times*, but since the philosophers in question think of times merely as spacetime regions of a special sort, this hedge isn't really needed.) Accordingly, friends of 3P often say that the third slot in parthood$_m$ can be filled by a spacetime region. This would make room for the possibility of an object that is a part$_m$ of a second object at one region but not at another.

In this paper I consider an alternative to Absoluteness and 3P, namely

Four-Place Parthood (4P): Parthood$_m$ is a four-place relation that can be expressed by the predicate 'x at w is a part$_m$ of y at z'.

It will be natural, though not officially required, for friends of 4P to say that parthood$_m$ has one slot for a part, a second slot for a *location* of that part (e.g., a spacetime region), a third slot for a whole, and a fourth slot for a *location* of that whole (e.g., a spacetime region). I will argue that much of the support for 3P is misplaced and should be redirected toward 4P.

---

4 'Exactly occupies' is usually left undefined. Informally, though, the idea is supposed to be that an object O exactly occupies a spacetime region R just in case O has (or has-at-R) precisely the same shape, size, and position as R. Thus big spheres exactly occupy only big spherical regions, etc. Moreover, there should be nothing obviously impossible about the claim that a thing exactly occupies each of two or more non-point-sized regions but not their sum or any of their proper subregions. It is worth noting that while this characterization focuses on spatiotemporal examples, it is not obvious that the only entities that can be exactly occupied are spacetime regions. Perhaps there are such things as (nonspatiotemporal) *argument places* in universals. If so, we may want to say that they are exactly occupied by various things. For skepticism about the intelligibility of 'exactly occupies' (and hence of the multi-location thesis), see Parsons (2008). For a defense, see Hudson (2008).

5 Hudson (2001: 61-71) endorses 3P, McDaniel (2004) claims that the defender of 'modal realism with overlap' should endorse it, and Donnelly (forthcoming) claims that multi-locationists more generally should endorse 3P (or some very similar thesis stated a bit differently).

Specifically, my main conclusion will be that anyone who accepts the thesis of multi-location mentioned above should prefer 4P to 3P.[6]

This is a diverse group. It includes *endurantists*[7] who say that a material object exactly occupies each in a series of temporally unextended 'slices' of its spacetime path (Mellor 1980, van Inwagen 1990, Rea 1998, Sattig 2006). But it also includes Hud Hudson (2001), who accepts a form of perdurantism[8] according to which ordinary material objects are multi-located 'spacetime worms' that exactly occupy many, mostly overlapping, four-dimensional spacetime regions. Finally, it includes the proponents of a certain position in the metaphysics of modality – namely, the Modal Realism with Overlap (MRO) set out by Kris McDaniel (2004). On this view, at least some material objects are 'wholly present' in many different concrete possible worlds; any such object would exactly occupy at least one different spacetime region for each of the worlds at which it exists (but the regions themselves are each confined to a single world). Though McDaniel himself neither accepts nor rejects MRO, he takes it to be a serious contender as a theory of de re modality and possible worlds.[9]

Multi-location, then, is a 'big tent' that includes a wide range of views about material objects. As I will understand it, however, the doctrine builds in quite specific and controversial assumptions about spacetime. In particular, it incorporates a form of spacetime substantivalism.[10] On this view, spacetime points and regions exist in their own right and are not to

6 I will ignore a fourth view about the adicity of parthood$_m$: viz., that it is a 'multigrade' or 'variably polyadic' relation, perhaps one that can hold between two objects (the 'part' and the 'whole') and various numbers of 'indices' such as times, places, spacetime regions, moments of proper or personal time, possible worlds, sortals, etc. I will assume that this is an answer of last resort to the Adicity Question.

7 Endurantism, roughly, is the view that material objects persist not by having different temporal parts at different times but by being wholly present at each time at which they exist.

8 Perdurantism, roughly, is the view that material objects persist by having different temporal parts at different times. This contrasts both with endurantism and with the 'stage view' (Sider 2001) according to which ordinary material objects are instantaneous stages that persist by bearing a temporal counterpart relation to other stages located at other times.

9 Some philosophers deny that material objects are multi-located but apparently accept (or, in the case of Lewis, remain agnostic about) multi-location for entities in other categories, such as universals (Newman (2002), Armstrong (1989), Lewis (1983, 1986a)) or tropes (Campbell (1981: 487), Ehring (1997: chs. 4 and 5)).

10 One might wish to combine the view that material objects are (in some sense) multi-located with a *relationist* theory of spacetime, according to which there are material objects and events standing in various spatiotemporal relations, but there are no spacetime points or regions. Perhaps one could hold that some material objects are multi-located in the sense of being at a spatial or temporal distance from themselves. Such a view does not count as a form of *multi-location*, as I will be using that term, and I will not consider the view any further here. Thanks to Kris McDaniel.

be reduced to things or events standing in spatiotemporal relations. Spacetime points are taken to be instantaneous, spatially unextended concrete particulars, and a spacetime region is taken to be any non-empty collection of such points. (Multi-location is neutral as to whether these 'collections' are *sets* that have their points as members or *sums* that have their points as parts.) Given this form of substantivalism, the best candidates for being *instants of time* are spacetime regions of a certain sort – viz., regions that are *temporally unextended*, so that any two points in such a region are simultaneous or spacelike separated, and *maximal,* i.e., not subregions of other temporally unextended regions. I will call such regions *global time-slices*.

So much for preliminaries. The plan for the rest of the paper is as follows. In section 2 I sketch a standard route from multi-location to the view that parthood$_m$ has more than two argument places. In section 3 I mount a detailed case against 3P, the view that parthood$_m$ is a three-place relation. In section 4 I show that if we shift to the view that parthood$_m$ is four-place, we can avoid the problems facing 3P, and I suggest that we do not face any comparably serious new problems. Finally, in section 5, I address some questions about how to construct a formal theory of parthood$_m$ on the assumption that it is a four-place relation.

2. From Multi-Location to the Denial of Absolutism

Let us say that a spacetime region R is a *location* of an object O just in case O exactly occupies R. To accept multi-location is to hold that at least some material objects have multiple locations. For those who endorse this view, it is natural to think that a multi-located object could exhibit a robust form of *mereological variation*: intuitively, it could be entirely made up of one collection of parts at one of its locations, while being entirely made up of a different collection of parts at another of its locations. Indeed, so far as I know, all actual multi-locationists do in fact embrace this sort of mereological variation. I will assume that if one takes on these commitments, then one should reject Absolutism.[11] I take this assumption to be quite widely held,[12] typically on the basis of something like the following line of thought.

Suppose that material object O exactly occupies region R1, and that a different material object, P, exactly occupies some proper subregion of

11 See Donnelly (forthcoming) for more on this issue.

12 One potential exception is Sattig (2006). Sattig endorses multi-location and mereological variation between locations, but I do not know whether he believes that there is a *fundamental* parthood relation that can hold between material objects.

R1. Then, in the absence of any considerations to the contrary, this makes it plausible that P bears a fundamental parthood relation to O (together perhaps with some region or regions).[13] That is, this makes it plausible that P is a *part$_m$* of O, perhaps *at* some region or regions. Further, suppose that O also exactly occupies some spacetime region R2 distinct from R1, but that P does not exactly occupy any subregion of R2. Then, again absent any considerations to the contrary, this makes it plausible that P bears the negation of that same parthood relation to O (together perhaps with some region or regions). In other words, this makes it plausible that P is a *non*-part$_m$ of O, perhaps at some region or regions.

Now, with all this in place, suppose that parthood$_m$ is two-place. Then the 'perhaps at some region or regions' clauses above never come into play, and we face pressure to say that P is both a part$_m$ of O (*simpliciter*) and a non-part$_m$ of O (*simpliciter*),[14] which is absurd. On the other hand, if parthood$_m$ has an extra argument place, we face no pressure to say this. Instead we can fall back on those 'perhaps at some region or regions' clauses: we can say that P is a part$_m$ of O at one region and a non-part$_m$ of

---

13 The main alternative is to claim that P merely bears some non-fundamental parthood relation to O, some relation that is defined in terms of a more natural parthood relation. Consider, e.g., the standard perdurantist treatment of the following case: an oxygen molecule enters my body, becomes 'caught up in my life' for a while, then leaves my body. Perdurantists typically hold that each object exactly occupies just one region – its entire 'spacetime path.' Since the oxygen molecule's path overlaps mine but is not a subregion of mine, the standard perdurantist will say that the molecule does not exactly occupy any subregion of any region that I exactly occupy. Given this view, we should deny that the molecule bears parthood$_m$ to me (anywhere). At best, it bears some *non*-fundamental parthood relation to me – e.g., a time-relative parthood relation defined in terms of parthood$_m$ as follows:

Part-at-t: x is part of y at t =df. (i) t is an instant of time and (ii) x's instantaneous temporal part at t is a part$_m$ of y's instantaneous temporal part at t.

Temporal part: x is an instantaneous temporal part at t of y =df. (i) t is an instant of time, (ii) x exists at t but only at t, (iii) x is a part$_m$ of y, and (iv) every part$_m$ of y that exists at t has a part$_m$ in common with x.

This allows the perdurantist to say that the oxygen molecule is a part of me at certain times despite not bearing any fundamental parthood relation to me. (This strategy is developed by Sider 2001.) This seems to be the best option for a pair of material objects neither of which exactly occupies any subregion of any region exactly occupied by the other. But given the fact that *our* object P *does* exactly occupy a proper subregion of a region that O exactly occupies, P and O seem to be especially good candidates for instantiating a *fundamental* parthood relation.

14 Even if parthood$_m$ is two-place, this conclusion can be avoided if some 'relativizing' theory is true of the instantiation relation, so that it has additional argument places for times or spacetime regions. (See Haslanger (2003) for a survey of 'relativizing' options vis-à-vis change with respect to apparently monadic intrinsic properties.) This alternative is much in the spirit of non-Absolutist theories of parthood$_m$. But the alternative theory makes it harder to set out a formal theory of parthood$_m$ than it is in the context of relativizing theories that put the extra argument place(s) into parthood$_m$ itself. See section 4.

O at a *different* region. Thus, if we think that material objects are multi-located and vary mereologically between locations, this will give us a quite powerful reason to postulate at least one extra argument place in parthood_m.

3. Problems for Three-Place Parthood

If parthood_m is a three-place relation, then one question that arises immediately is what I will call the *Restriction Question*: what conditions does a spacetime region R have to meet in order for a material object x to be a part_m of a material object y *at* R? Must R be a maximal spatiotemporally interrelated region[15] – i.e., a 'complete spacetime'? Must R be a global time-slice of a spacetime – roughly, an instant? Must R be a location of x? A location of y? Must R overlap some location of one or both of them? I will argue that the Restriction Question gives rise to serious problems for 3P, and that 4P avoids these problems entirely.

I will begin by describing a very simple case of parthood_m. Concerning this case, I will ask: relative to which region or regions does parthood_m hold? I examine what I take to be the best answers available to the defender of 3P, and I argue that each of them has major drawbacks. I then show that if we reject 3P in favor of 4P, we can answer the Restriction Question in a way that does not give rise to any comparably serious objections.

Here is the case, which I will call *Case 1*. Material objects a and b are each 'mono-located': each of them exactly occupies just a single spacetime region. Object a's location is the temporally unextended region Ra, and b's location is the temporally unextended region Rb, which does not overlap with Ra. Material object c is also mono-located: it exactly occupies Rc, which is the sum or union of Ra and Rb. Intuitively, we should think of a and b as being non-overlapping proper parts of c, and we should think of c as having no parts that are disjoint from each of a and b.

15 To say that R is a maximal spatiotemporally interrelated spacetime region is to say that (i) R is a spacetime region, (ii) each point in R is spatiotemporally related to each other point in R, and (iii) no point that is not in R is spatiotemporally related to any point in R.

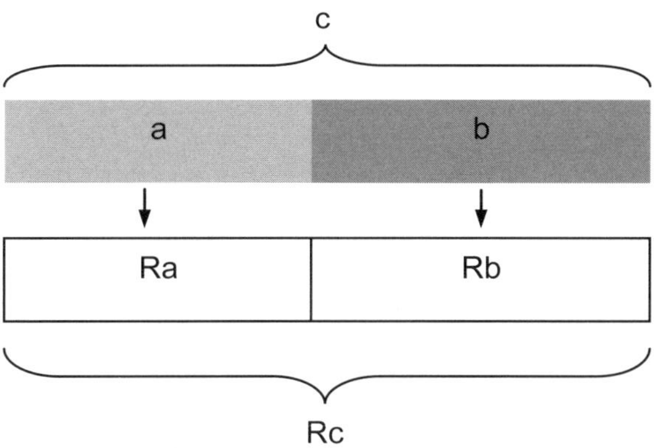

In Case 1, it is plausible that a is a part$_m$ of c 'exactly once'. It seems that there is just one 'instance' of parthood$_m$ involving a and c.

Figure 1

To avoid prejudging the main question, however, we make no stipulations as to *where* a or b are parts$_m$ of c. Instead, we can consider a list of candidates for being regions at which a is a part$_m$ of c. (For simplicity we ignore b.)

Ra:	a's location.
Rc:	c's location.
Rmax:	a maximal spatiotemporally interrelated region (a 'complete spacetime') that has Rc as a proper subregion.
Rblob:	a four-dimensional region that has Rc as a proper subregion and Rmax as a proper superregion.
Rslice:	a global time-slice that has Rc as a proper subregion.
Rslice-:	an instantaneous region that has Rc as a proper subregion and Rslice as a proper superregion.
Rchunk:	a global 'time-chunk' of a certain temporal thickness that has Rslice as a proper subregion and Rmax as a proper superregion.
Ra+:	a region that is a proper superregion of Ra and a proper subregion of Rc.
Ra-:	a proper subregion of Ra.
Rc-:	a proper subregion of Rc that does not overlap Ra.
Re:	a subregion of Rmax that does not overlap Rc.

At which of these regions, if any, is a a part$_m$ of c? Two types of answers are available to proponents of 3P: pluralist answers, according to which a is a part$_m$ of c at more than one of the given regions, and non-pluralist answers, according to which a is a $part_m$ of c at no more than one of the given regions.

3.1 Pluralist Answers

Pluralist answers are vulnerable to a simple but compelling objection.[16] Intuitively, a is a $part_m$ of c 'only once'. Relatedly, there seems to be just one 'case' or 'instance' of $parthood_m$ involving a and c. But if $parthood_m$ were three-place and a were a $part_m$ of c at multiple regions, then there would be *multiple* cases or instances of $parthood_m$ involving those objects. So, given 3P, the pluralist answers are false. It will be convenient to set this out in numbered form:

P1: If $parthood_m$ is three-place and a is a $part_m$ of c at more than one of the given regions, then there is more than one instance of $parthood_m$ involving a and c.
P2: There is exactly one instance of $parthood_m$ involving a and c.
C: So, if $parthood_m$ is three-place, then a is a $part_m$ of c at no more than one of the given regions.

Let me begin by saying a bit more about P1. I assume that, regardless of what one ultimately wants to say about the metaphysics of property instances (and relation instances),[17] and regardless of one's favored answer to the Adicity Question or the Restriction Question, one can agree that the number of instances of $parthood_m$ is equal to the number of ordered n-tuples whose members instantiate $parthood_m$ (in the order given by the n-tuple).[18] More specifically, I assume that for any number, #, there are # instances of $parthood_m$ involving a and c just in case there are # n-tuples containing a and c whose members instantiate $parthood_m$ (in the order given by the n-tuple).

16 As I note later, these answers are also vulnerable to problems arising from the 'two case intuition' to be discussed in 3.2.3.

17 One might hold that an instance of an n adic relation R is merely an ordered pair $\langle R, \langle o_1 \ldots o_n \rangle\rangle$ such that the members of $\langle o_1 \ldots o_n \rangle$ instantiate R in the given order.

18 One can accept this claim without accepting a generalized variant of it to the effect that for *any* relation R, the number of instances of R = the number of n-tuples whose members instantiate R (in the given order). One might think that symmetric relations, such as being two feet away from, are counterexamples to the more general principle, for one might think that in a situation in which a is two feet from b, there are two ordered pairs, $\langle a, b \rangle$ and $\langle b, a \rangle$, whose members instantiate the relation in the given order, but one might find it counterintuitive to say that in such a situation we have two instances of being two feet away from. Nevertheless, since $parthood_m$ is not symmetric, this style of case does nothing to cast doubt on the original claim stated in the main text.

Now, if parthood$_m$ were three-place, and a were a part$_m$ of c at multiple regions – say, at exactly two distinct regions, r and r* – then there would be multiple ordered triples containing a and c whose members instantiate parthood$_m$ (in the order given by the triple): namely, ⟨a, c, r⟩ and ⟨a, c, r*⟩. But in that case, there would be *two* instances of parthood$_m$ involving a and c. This confirms P1.

We can turn now to P2, which I will call the *single case intuition.* It says that we have just a single instance of parthood$_m$ involving a and c. This claim can be motivated in at least three ways. First and most importantly, it should seem highly plausible on its own. It is not the sort of thing that anyone would even think to *question* had it not been used as a premise in an argument. Suppose that, in another context, we were given the set-up of Case 1, together with its depiction in Fig. 1, and we were asked, "How many instances of parthood$_m$ involving the objects a and c do we have in this case?" We would not hesitate to answer, "One," or so I conjecture.

Moreover, if the set-up of the case had been *different* in certain ways, we would have had different intuitions about how many such instances the case contains. Suppose, e.g., that c had *two* nonoverlapping locations, and that each of these contained a location of a as a subregion. (See Fig. 2.) Then (other things being equal) it would seem appropriate to say that a was a part$_m$ of c 'twice' – once above and once below – and that, correspondingly, there were *two* instances of parthood$_m$ involving a and c.

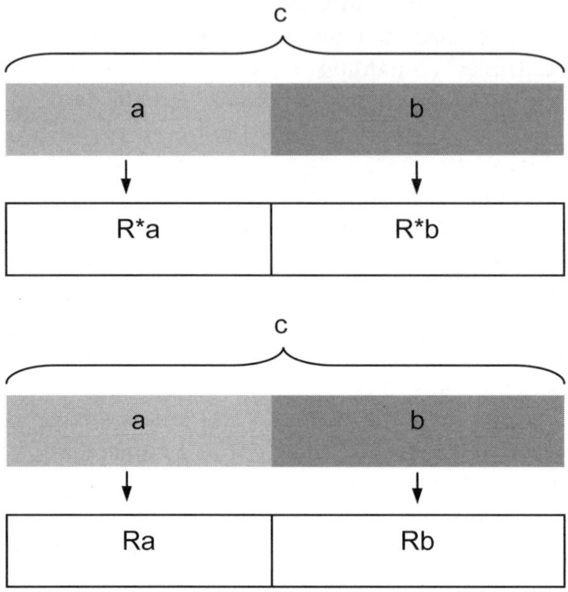

In this *variant* of Case 1, it is plausible that a is a part_m of c 'twice'. It seems that there are two 'instances' of parthood_m involving a and c, one that is associated with the lower portion of the diagram, and another that is associated with the upper portion of the diagram.

Figure 2

But as things stand in Case 1, c has just one location, and a has just one location (a subregion of c's). Someone who was willing to speak of 'occurrences' of material objects would say that we have just one occurrence of c and just one occurrence of a, and that the latter is a part *simpliciter* of the former. In these circumstances, it is overwhelmingly plausible that a is part_m of c just once, and that there is just one instance of parthood_m involving the two objects.

A *second* line of motivation for the single case intuition looks to the verdicts of Absolutism. The thought here is that Absolutism, in virtue of its simplicity and its pedigree[19], has a kind of default status, so that if certain pieces of data (e.g., multi-location together with mereological variation between locations) force us to *reject* Absolutism in favour of 3P or 4P, then we should depart from Absolutism *only as much as is necessary to accom-*

19 Traditional formal theories of the part-whole relation employ a two-place parthood predicate. See Simons (1987) for a survey. Contemporary perdurantists (with the exception of Hudson (2001)) all speak as though they accept Absolutism, as do contemporary *presentists* (who hold that only what is present exists at all). Moreover, even some multi-locationists (van Inwagen 1990) find it convenient to work under the pretense that parthood_m is two-place.

modate the data in question. In particular, we should strive to make our non-Absolutist theory of parthood$_m$ resemble Absolutism as closely as possible in important respects (while still accommodating the data). One important respect in which two theories of parthood$_m$ can agree concerns their verdicts on the single case intuition.

Absolutism delivers a clear verdict in favor of that intuition. According to Absolutism, a and c belong to just one ordered n-tuple whose members instantiate parthood$_m$ (in the order specified by the n-tuple) – namely, the ordered pair ⟨a, c⟩. Absolutists, therefore, will say that in Case 1 we have just a single instance of parthood$_m$ involving a and c. Other things being equal, then, non-Absolutists should prefer a theory that lets them say this as well.

A *third* motivation for the single case intuition appeals to considerations of parsimony. For properties and relations as fundamental as parthood$_m$, we should seek to avoid redundancy: other things being equal, we should hold that these properties and relations are instantiated only as many times as is needed to fully account for how things are. It seems clear that in Case 1, we can fully account for how things are – and in particular, we can fully account for the mereological relationship between a and c – without claiming that a and c instantiate parthood$_m$ many times over (at multiple regions). On the contrary, that claim would result in something analogous to an *over-determination* of their mereological relationship; it 'over-characterizes' this relationship. To avoid this sort of redundancy, we should embrace the single case intuition.

It is worth noting that the single case intuition is perfectly consistent with the view that there are *non*fundamental parthood relations that hold more than once between a and c. Consider the relation defined as follows:

D3: x is a part of y within R =df. there is some spacetime region R* such that x is a part$_m$ of y at R* and R* is a subregion of R.

If a is a part$_m$ of c at a region R*, then a is a part of c *within* every superregion of R*. So unless R* has no proper superregions, the relation of part-within will hold between a and c many times over. But this does nothing to undermine the thought that the *fundamental* parthood relation for material objects, parthood$_m$, holds between a and c just once.

In sum, then, the situation is this. In order to respect the single case intuition, the friend of 3P must deny that a is a part$_m$ of c at more than one region: he should reject the pluralist answers to our question about a and c.[20]

3.2 Non-pluralist Answers

Non-pluralist answers all say that a is a $part_m$ of c at no more than one of the regions listed earlier. Interestingly, the single case intuition helps us to dispense with several of these answers as well.

For example, we can immediately reject the suggestion that a is not a $part_m$ of c at any of those regions. Given the single case intuition and 3P, a must be a $part_m$ of c at some region. But surely we haven't excluded all of the best candidates from our list! If a is $part_m$ of c at any region, surely it is a $part_m$ of c at *some* region on the list.

Likewise, we can reject the view that a is a $part_m$ of c at exactly one of the following: Rblob, Rchunk, Rslice-, Ra+, Ra-, Rc-, or Re. Suppose, for example, that a is a $part_m$ of c at Rblob. Then, since there is nothing special about that region, presumably a must also be a $part_m$ of c at all such regions – i.e., at all four-dimensional regions that are both proper superregions of Rc and proper subregions of Rmax. To say that a is a $part_m$ of c at Rblob but not at any of these other regions like it would be unacceptably arbitrary. Relations as metaphysically basic as $parthood_m$ are just not

20 Some multi-locationists may be tempted to deny the possibility of Case 1 as a way of resisting my argument against the pluralist answers. Consider first the endurantist who denies the possibility of instantaneous material objects, such as a and b. Any material object, he says, must have a temporally extended path and must be multi-located within that path. In response, I say that whatever (slight) plausibility this may have as applied to material simples, it has even less as applied to composite material objects. Suppose that a itself is simple and has a temporally extended path whose final slice is Ra, and that b is also simple and has a temporally extended path whose first slice is Rb. Thus a and b coexist for just a single instant, and at that instant they compose the instantaneous composite object, c. As before, we have the intuition that a is a $part_m$ of c just once.

Perhaps some multi-locationists will go so far as to deny the possibility of instantaneous *composite* material objects as well. But then we can note that even if a is a part of c throughout some extended interval (so that a is $part_m$ of c at different regions associated with different global time-slices), we retain the intuition that there is at most one instance of $parthood_m$ involving a, c, *and Rslice or any of its subregions.* So, given 3P, we should not say that a is $part_m$ of c at each of the following: Ra, Rc, Rslice-, and Rslice. Rather, given 3P, we should say that a is $part_m$ of c at no more than one of those regions.

Similarly, perhaps some defenders of Modal Realism with Overlap will deny the possibility of material objects that exist in just one concrete possible world. But even if we grant that a and c have other locations in other worlds, we retain the intuition that a is $part_m$ of c just once *in the original world*: i.e., there is at most one instance of $parthood_m$ involving a, c, and Rmax or any of its subregions. (According to MRO as McDaniel presents it, spacetime regions, unlike material objects, are worldbound.) Together with 3P, this generates pressure to say that a is a $part_m$ of c at no more than one of these regions.

that haphazard. Given the single case intuition, however, we cannot say that a is a $part_m$ of c at *all* of the many regions like Rblob. So we should deny that a is a $part_m$ of c at Rblob. Parallel arguments can be given to show that a is not a $part_m$ of c at Rchunk, Rslice-, Ra-, Rc-, or Re. None of these regions can be plausibly said to be the *only* region at which a is a $part_m$ of c. This leaves us with four remaining candidates: Rmax, Rslice, Ra, and Rc.

3.2.1 The Spacetime Principle

We can begin by considering the *Rmax view*, according to which a is a $part_m$ of c at Rmax and only there. (Rmax, recall, is the complete spacetime in which a and c are embedded.) This view is associated with the following general principle:

The Spacetime Principle: Necessarily, for any material objects x and y and any z, if x is a $part_m$ of y at z, then z is a maximal spatiotemporally interrelated spacetime region (a 'complete spacetime') that includes a location of x and a location of y as subregions.[21]

I will assume that if the Rmax view is true, then the Spacetime Principle is true as well. Although this assumption could be questioned, it will appeal to those who agree that properties and relations as metaphysically basic as $parthood_m$ are not haphazard. $Parthood_m$, for example, does not hold relative to a complete spacetime in one case and relative to an arbitrarily selected proper subregion of a spacetime in another, otherwise similar case. Rather, properties and relations as basic as $parthood_m$ are governed by relatively simple general rules and have their 'core features', such as their adicities, essentially. I take it that if $parthood_m$ is in fact governed by the Spacetime Principle, then this is one of its core features. (Likewise for the competing general principles that I discuss.)

The Spacetime Principle might appeal to a certain subset of those philosophers who endorse MRO. As I mentioned earlier, MRO is a form of realism about possible worlds according to which (i) possible worlds are

21 To this we might plausibly add some further clause to the effect that x's location in z is a subregion of y's location in z. It will be unclear how this further clause should be specified if one believes that some objects are multi-located within a single spacetime, but as I note below, anyone who believes this is unlikely to accept the Spacetime Principle. See Donnelly (forthcoming) for a discussion of related questions.

concrete entities (specifically, they are 'complete spacetimes') and (ii) at least some material objects are 'wholly present' in more than one world, where being wholly present in a given world is understood as exactly occupying some subregion of that world. Friends of MRO will disagree amongst themselves as to whether material objects ever exactly occupy multiple subregions of a given world. Those who endorse multi-location within worlds will reject the Spacetime Principle,[22] but those who reject multi-location within worlds may initially be more sympathetic to that principle.[23]

The Rmax view and the Spacetime Principle mesh well with the single case intuition, and thus they avoid the problems facing the previous answers that we have considered. Together with the set-up of Case 1, the Rmax view entails that there is just one region at which a is a $part_m$ of c, and hence that a is a $part_m$ of c 'just once'. Moreover, there is nothing arbitrary about a's being a $part_m$ of c only at Rmax. After all, Rmax is quite special as far as a and c are concerned: it is the only complete spacetime that contains either of them.

Still, the Spacetime Principle has a drawback. To state the argument against this principle, I will need to invoke some technical terminology. I take it that the notion of intrinsicness, as applied to properties, is familiar and well understood (even if its precise definition remains elusive). Roughly, a property P is intrinsic just in case whether or not an object O has P depends only on what O is like in itself and is independent of how O is related to things separate from itself. Or, in different terms, a property is intrinsic iff it cannot differ between duplicates.

22 These philosophers will presumably want to say that material objects often vary mereologically between locations within a single spacetime – e.g., that at one of my locations within this spacetime, I am made up of one collection of particles, and at another of my locations within this same spacetime, I am made up of a different collection of particles. On this view, there will be many true sentences of the form, 'Material object p is a $part_m$ of material object o at region r1 but not at region r2, although r1 and r2 are both proper subregions of the same spacetime'. This obviously conflicts with the Spacetime Principle.

23 According to them, mereological variation between locations never occurs within a single spacetime. If I exactly occupy R and R*, and if, 'as I am at R', I have a certain skin cell as a $part_m$ but, 'as I am at R*', I do not have that cell as a $part_m$, then these locations will be subregions of different spacetimes – say, w and w*. This will allow us to account for any mereological variation between my locations by relativizing to spacetimes. We will be able to say: the cell is a $part_m$ of me at spacetime w but is not a $part_m$ of me at spacetime w*. Granted, mereological variation *between locations* is not the only sort of mereological variation (or apparent mereological variation) that these philosophers will need to account for. Such variation might also occur *over time* but within a single location. There is at least a sense in which, e.g., a certain hair, h, is a part of me at t but not at t*, after it falls out. The MROists in question will see this as variation *within a single location*. On the assumption that they posit temporal parts, they can describe the case by saying that the t-part of h is a $part_m$ of my t-part at w (this spacetime) while the t*-part of h is not a $part_m$ of my t*-part at w.

It is also fairly easy to get a grip on a notion of intrinsicness for *relations*, although the terminology here is a bit less familiar. We can start with the notion of *plural duplication*. (I take this term from McDaniel 2008.) Intuitively, to say that the Xs and the Ys are plural duplicates is to say that the Xs match the Ys with respect to their intrinsic properties and internal arrangement, though not necessarily with respect to how they are related to outside things.

For example, suppose that Tom, Dick, and Harry are standing in a column, single-file, all facing North, and are arranged by increasing height, so that Tom, the shortest, is in front, and Harry, the tallest, is in the back, with each of them being exactly two feet away from Dick, who is in the middle. Further, suppose that Tom*, Dick*, and Harry* are intrinsic duplicates of Tom, Dick, and Harry, respectively, and are arranged in an exactly similar fashion with regard to their spatial and causal relations. Then Tom, Dick, and Harry are plural duplicates of Tom*, Dick*, and Harry*, even if the latter trio are, say, facing in a different direction, or nearer to a lake, than are the former trio.

If we help ourselves to the notion of an intrinsic property and to Lewis's notion of a *perfectly natural* relation (see note 2), then we can offer a more formal definition of plural duplication as follows:

D4: the Xs and the Ys are plural duplicates =df. there is a one-one correspondence between the Xs and the Ys that preserves intrinsic properties and perfectly natural relations – i.e., a bijective function f such that (i) for any intrinsic property P and for any x among the Xs, x instantiates P iff f(x) instantiates P, and (ii) for any perfectly natural relation R and any ordered n-tuple $\langle x_1 \ldots x_n \rangle$ of the Xs, $x_1 \ldots x_n$ instantiate R in that order iff $f(x_1) \ldots f(x_n)$ instantiate R in that order.

With the notion of plural duplication in hand, I can define an *intrinsic relation* as one that cannot differ between pluralities that are plural duplicates. Thus, if the Xs and the Ys are plural duplicates, if R is an intrinsic relation, and if the Xs instantiate R (in some order), then the Ys instantiate R as well (in some order). I assume that being larger than and being two feet apart are both intrinsic relations and that having a common owner and being two feet apart and within three miles of a lake are *extrinsic* (non-intrinsic) relations.[24]

24 Lewis (1986a: 62) offers a very closely related set of definitions. He defines an *internal* relation as one that 'supervenes on the intrinsic natures of its relata', and he defines an *external* relation as one that is not internal but nevertheless 'supervenes on the intrinsic nature of the composite of the relata taken together'. Bricker (1993: 274) systematizes the termi-

My argument against the Spacetime Principle hinges upon one crucial assumption: that parthood$_m$, whatever its adicity, is an intrinsic relation. I will call this the *intrinsicness intuition*, and I will try to show that it cannot be plausibly combined with the package consisting of the Rmax view and the Spacetime Principle.

Let me begin by trying to motivate the intrinsicness intuition, albeit briefly. First, one might reasonably take parthood$_m$, or indeed any fundamental parthood relation, to be a perfectly natural relation, in which case the definitions of plural duplication and 'intrinsic relation' given above will guarantee that parthood$_m$ is intrinsic. Second, one might once again appeal to the principle that non-Absolutist theories of parthood$_m$ should depart from Absolutism *only as much as is necessary to accommodate the relevant data.* In that case, since parthood$_m$ is plausibly an intrinsic relation according to Absolutism, the friend of 3P or 4P should strive to accommodate the intrinsicness intuition too.

Third and most importantly, I take the intrinsicness intuition to be highly plausible on its own, even in the absence of any supporting argument. Suppose that material object p is a part$_m$ of material object o at spacetime region r, and that p*, o*, and r* are plural duplicates of p, o, and r (respectively[25]). This means that p* is a duplicate of p, o* is a duplicate of o, and r* is a duplicate of r; and that p*, o*, and r* are 'inter-related amongst themselves' in exactly the same manner as are p, o, and r. Might it be that whereas p *is* a part$_m$ of o at r, p* is *not* a part$_m$ of o* at r*, despite all these similarities between the two trios? For those who find this as hard to believe as I do, the intrinsicness intuition will be compelling.

Now, to see why this intuition conflicts with the package consisting of the Rmax view and the Spacetime Principle, consider the trio of a, c, and Rc, the last of which is c's location. According to the Rmax view, the trio do not instantiate parthood$_m$: a is a part$_m$ of c at Rmax, not at Rc. But presumably there can be a trio, a*, c*, and Rc*, that are plural duplicates of a, c, and Rc (respectively), and that *do* instantiate parthood$_m$. Just suppose that Rc* is a complete spacetime, but that a*, c*, and Rc* (and their

nology (in a manner implicit in various things Lewis says) by defining an *intrinsic* relation as one that is either internal or external, and an *extrinsic* relation as one that is not intrinsic. Lewis takes being larger than to be internal (hence intrinsic, in Bricker's terminology), being two feet away from to be external (hence intrinsic, in Bricker's terminology), and having a common owner to be neither internal nor external (hence extrinsic, in Bricker's terminology).

25 To say that $o_1 \ldots o_n$ are plural duplicates of $o^*_1 \ldots o^*_n$ respectively is to say that the function f that pairs o_1 with o^*_1 and . . . and o_n with o^*_n preserves intrinsic properties and perfectly natural relations. Hence if $o_1 \ldots o_n$ are plural duplicates of $o^*_1 \ldots o^*_n$ respectively, and R is an intrinsic relation, and $o_1 \ldots o_n$ instantiate R in that order, then $o^*_1 \ldots o^*_n$ instantiate R in that order as well.

internal arrangement) are as much like a, c, and Rc (and their internal arrangement) as is compatible with this difference. Thus a* is a duplicate of a, c* is a duplicate of c, and Rc* is a duplicate of Rc. Moreover, the spatiotemporal and causal relations between a*, c*, and Rc* match those between a, c, and Rc: the two trios are 'inter-related amongst themselves' in exactly the same manner.

The big difference between the two trios, of course, is that the original trio (of a, c, and Rc) are embedded within a larger spacetime, so that Rc is a mere proper subregion of a spacetime, rather than being a complete spacetime in itself. The new trio (of a*, c*, and Rc*) do not find themselves embedded in a larger spacetime; Rc* is not a subregion of any larger region. This difference, however, is clearly an *extrinsic* difference: it concerns only the ways in which the members of a trio are related to things outside of themselves. Intrinsically, the two trios are just alike. They are plural duplicates.

Given the Spacetime Principle, we are forced to conclude that a* is a $part_m$ of c* at Rc*: after all, clearly a* is a $part_m$ of c* at *some* region, and the only spacetime that has locations of a* and c* as subregions is Rc*. So, if the Spacetime Principle and the Rmax view are true, then a* is a $part_m$ of c* at Rc*, whereas a is not a $part_m$ of c at Rc, despite the fact that the two trios are plural duplicates. But if $parthood_m$ could differ in this way between plural duplicates, it would not be an intrinsic relation, contrary to the intrinsicness intuition. Call this style of argument a *contraction* argument.[26]

3.2.2 The Time-Slice Principle

Endurantists sometimes say that $parthood_m$ is a 'time relative relation'. On one natural interpretation, this means that $parthood_m$ is a three-place relation that can hold between two material objects and an instant of time. Some endurantists seem to accept this and, further, to endorse *multi-location*, according to which instants are spacetime regions of a special sort (global time slices).

26 A second argument against the Spacetime Principle arises from the possibility of what Parsons (2007) calls 'knuggy' worlds, worlds in which there are spacetime regions, and each of them has another such region as a proper superregion. Such worlds contain no maximal regions and, assuming that each point in such a world is spatiotemporally related to every other such point, they contain no maximal spatiotemporally interrelated regions – i.e., no 'complete spacetimes'. On the assumption that $parthood_m$ can be instantiated within such worlds, the Spacetime Principle must be rejected.

With regard to Case 1, these endurantists will presumably want to endorse the *Rslice view*, according to which a is a part$_m$ of c at exactly one region: a global time-slice that has Rc as a proper subregion.[27] But it turns out that the Rslice view conflicts with the intrinsicness intuition just as badly as the Rmax view does. For the Rslice view, too, is vulnerable to a contraction argument.[28]

As in the case of the Rmax view, I take it that the Rslice view is plausible only if accompanied by a corresponding general principle, which in this case would be

The Time-Slice Principle: Necessarily, for any material objects x and y and any z, if x is a part$_m$ of y at z then z is a global time-slice that has a location of x and a location of y as subregions.[29]

As applied to the 'contracted' case of a*, c*, and Rc* discussed earlier, the Time-Slice Principle will force us to conclude that a* is a part$_m$ of c* at Rc*. When this conclusion is combined with the Rslice view and its claim that a is *not* a part$_m$ of c at Rc (despite the fact that the trios are plural duplicates), we again get the unappealing result that parthood$_m$ is not an intrinsic relation. This shows that the package consisting of the Rslice view and the Time-Slice Principle also conflicts with the intrinsicness intuition.

The Time-Slice Principle should be distinguished from

The Simple Instants Principle: Necessarily, for any material objects x and y and any z, if x is a part$_m$ of y at z then z is a simple, *sui generis* instant of time and x and y both exist at z.

27 If the spacetime in Case 1 is relativistic, there might be many different global time-slices (i.e., many different maximal spacelike hypersurfaces) that have Rc as a subregion. Again, the single case intuition will put pressure on us to say that that a is a part$_m$ of c at no more than one of these global time-slices.

28 Some of these philosophers may have been tempted to say that parthood$_m$ holds relative not merely to instants but also to temporally extended intervals. Given spacetime substantivalism, this would lead naturally to the view that a is a part$_m$ of c not just at a global time-*slice* but also at the many global time-*chunks* (e.g., Rchunk) that include their locations as subregions. This package conflicts with the single case intuition discussed in section 3.1. It is also vulnerable to a contraction argument, and so conflicts with the intrinsicness intuition, just as does the Time-Slice Principle + Rslice view package.

29 We might add some clause to the effect that x's location in z is a subregion of y's location in z. It becomes more difficult to specify the clause if we want to allow for the possibility that x and/or y have multiple locations within the same time-slice. See Donnelly (forthcoming) for more on this issue.

This principle is most plausible in the context of the view that (i) space is a (presumably three-dimensional) manifold whose basic constituents are simple, enduring spatial points and (ii) time is a separate (presumably one-dimensional) manifold whose basic constituents are simple, non-persisting instants. This view about space and time conflicts with multi-location.

The Simple Instants Principle (SIP) is apparently invulnerable to contraction arguments. To apply such an argument to it, we would need to find 'plurally duplicate ordered triples', ⟨p, o, t⟩ and ⟨p*, o*, t*⟩ only one of which passes SIP's test for parthood_m. These triples would need to be such that (i) p, o, p*, and o* are all material objects, (ii) t is a simple instant and both p and o exist at t, but (iii) either t* is not a simple instant or either p* or o* fails to exist at t*. But since being a simple, *sui generis* instant is plausibly an intrinsic property, and since existing at is plausibly an intrinsic relation, it seems that we will not be able to find plurally duplicate ordered triples that differ in this way. (To say that the ordered n-tuples $\langle o_1 \ldots o_n\rangle$ and $\langle o^*_1 \ldots o^*_n\rangle$ are 'plurally duplicate ordered n-tuples' is to say that the bijection f that pairs the ith member of the first n-tuple with the ith member of the second preserves intrinsic properties and perfectly natural relations, and hence preserves intrinsic relations more generally.)[30]

So the doctrine of 'three-place, temporally relativized parthood_m' may be perfectly tenable in a context that includes simple, *sui generis* instants, but the doctrine becomes quite implausible when instants are treated merely as global time-slices, as they must be given multi-location.[31]

3.2.3 The Whole Location Principle and the Part Location Principle

Return to the question, "at which region or regions is a a part_m of c?" Two answers remain. According to the *Ra view*, a is a part_m of c at a's location, Ra, and only there. According to the *Rc view*, a is a part_m of c at c's location, Rc, and only there. Each of these answers is associated with a general principle. The Ra view corresponds to

30 Likewise for the view that parthood_m can hold relative to temporally extended intervals composed of simple, *sui generis* instants. Since the property being a temporally extended interval composed of simple *sui generis* instants is plausibly intrinsic, the view in question will not be vulnerable to contraction arguments.

31 A second objection to the Time-Slice Principle arises from the possibility of spacetimes that do not contain global time-slices. (Such spacetimes are permitted by General Relativity. For discussion, see Earman (1995).) On the assumption that parthood_m can be instantiated within these spacetimes, the Time-Slice Principle must be rejected.

The Part Location Principle: Necessarily, for any material objects x and y and any z, if x is a $part_m$ of y at z, then z is a spacetime region and x exactly occupies z.[32]

Likewise, the Rc view corresponds to

The Whole Location Principle: Necessarily, for any material objects x and y and any z, if x is a $part_m$ of y at z, then z is a spacetime region and y exactly occupies z.[33]

I assume that neither the Ra view nor the Rc view is plausible unless accompanied by the appropriate general principle. That is, I treat each view as a component of a larger, more general package.

Both packages fare quite well with respect to the considerations that we have discussed so far. We can begin with the single case intuition. There is nothing wildly implausible about the view that Ra is the *only* region at which a is a $part_m$ of c, nor about the view that Rc is the only such region. Thus neither of those regions is obviously *non-special* in the way that Rblob is. If, for example, we endorsed the Rc view, we could not be accused of thereby making $parthood_m$ seem excessively haphazard. (Though perhaps there is still something a *bit* arbitrary about preferring either the Part Location or the Whole Location Principle to the other.)

As for the intrinsicness intuition, neither of the relevant packages is vulnerable to a contraction argument. The problem with the Time-Slice Principle is that it sometimes treats plurally duplicate trios differently, forbidding one from instantiating $parthood_m$ but not forbidding the other. The Spacetime Principle also does this. This makes those principles (or the associated packages) vulnerable to contraction arguments. But neither the Whole Location Principle nor the Part Location Principle ever treats plurally duplicate trios differently, and so they are invulnerable to contraction arguments.

Consider the Whole Location Principle (WLP). To apply such a contraction argument to it, we would need to find 'plurally duplicate ordered triples', ⟨p, o, r⟩ and ⟨p*, o*, r*⟩, such that the former but not the latter satisfies WLP's necessary condition on $parthood_m$. These triples would need to be such that (i) p, o, p*, and o* are all material objects, (ii) r is a spacetime region that o exactly occupies, but (iii) either r* is not a space-

32 We might add ". . . and y exactly occupies some superregion of z." I suspect that anyone who finds the Part Location Principle plausible will find the stronger principle roughly equally plausible.

33 We might add ". . . and x exactly occupies some subregion of z."

time region or o* does not exactly occupy r*. Since being a spacetime region is plausibly an intrinsic property, and since exact occupation is plausibly an intrinsic relation, it seems that we will not be able to find plurally duplicate ordered triples that differ in this way. Parallel remarks apply, *mutatis mutandis*, to the Part Location Principle. This shows that both principles respect the intrinsicness intuition.

Despite these virtues, both principles have defects. Again we can start with the Rc view and the Whole Location Principle. The main problem for this package arises from questions about how to combine it with the widely accepted view that parthood$_m$ is governed by a *transitivity principle*, or at least by some straightforward analogue of such a principle.

Strictly speaking, of course, transitivity can be a property of two-place relations only. Thus if we insist that parthood$_m$ must turn out to be transitive in the strictest possible sense, we should cling to Absolutism; only Absolutists can take parthood$_m$ (expressed by ‘$<$’) to be governed by

Transitivity2P: $\forall x \forall y \forall z[(x<y \ \& \ y<z) \rightarrow x<z]$

However it is often noted that there is a very natural and straightforward *analogue* of the transitivity principle that presumably governs parthood$_m$ if that relation has three argument places. If we symbolize the predicate ‘x is a part$_m$ of y at z’ as ‘$x<_z y$’, then the analogue is:

Transitivity3P: $\forall x \forall y \forall z \forall w[(x<_w y \ \& \ y<_w z) \rightarrow x<_w z]$

In words, this says that if x is part$_m$ of y at w and y is part$_m$ of z at w then x is part$_m$ of z at w. A somewhat different way of capturing the intuitive idea underlying this principle is to say that the three-place relation part$_m$-of-at is such that for any r, the *two*-place, ‘indexed’ relation part$_m$-of-at-r is transitive in the strict sense.

There are, of course, many other vaguely ‘transitivity-like’ principles that can be framed in terms of a three-place parthood predicate. For example, there is the principle that if x is a part of y at some region and y is a part of z at some (perhaps different) region, then x is a part of z at *some* region. But I take it that, of these principles, Transitivity3P is *by far* the most natural analogue of Transitivity2P.

Return now to the Rc view and the Whole Location Principle. The problem for this package is not that it forces us to *reject* Transitivity3P; it doesn't. The problem is that the given package makes the principle *impotent* or *inapplicable* in certain contexts in which, intuitively, we should be able to use the principle to prove a certain fairly salient conclusion.

To see this, let us further specify Case 1, involving a, b, and c. Suppose that there is an additional singly located object, b*, and at least one additional composite material object, d, which is also singly located, and which we can think of as being composed of c and b*. Let spacetime region *Rb** be the location of b*, and let spacetime region *Rd* be d's location. We can suppose that Rb* does not overlap Rc and that Rd is the sum or union of these two regions. Finally, we can suppose that Rd is a proper subregion of Rslice and hence is temporally unextended. Thus the case can be represented by Fig. 3:

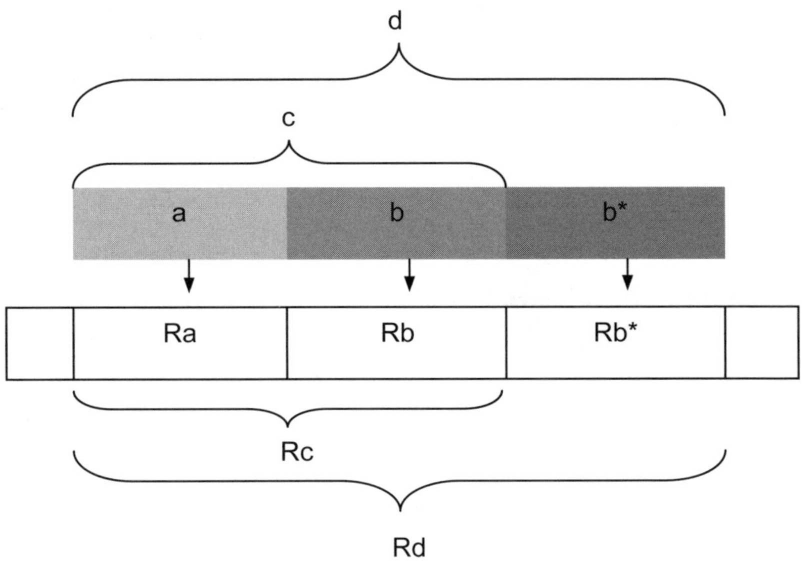

Figure 3

Intuitively, this is a case in which the transitivity of parthood$_m$ (or the most natural appropriate analogue of it) should be applicable. We should be able to combine that principle with a premise to the effect that a is a part$_m$ of c (at a certain region or regions, perhaps) and a premise to the effect that c is a part$_m$ of d (at a certain region or regions, perhaps) to yield a conclusion to the effect that a is a part$_m$ of d (at a certain region or regions perhaps). Somewhat more precisely, we should be able to formulate a three-premise argument that is formally valid in standard predicate logic and that has the following features:

(i) Its first premise is a true atomic sentence to the effect that a is a part$_m$ of c, at whatever region or regions (if any) participate in the relevant 'instance' of parthood$_m$. Thus, if parthood$_m$ is a two-place relation, then the first premise of the argument contains a two-place predicate (expressing parthood$_m$) attached to names for a and c: for example, "$a<c$" or "Pac". And if parthood$_m$ is a three-place relation, then the first premise contains a three-place predicate (expressing parthood$_m$) attached to names for a, c, and whatever region it is at which a is a part$_m$ of c: for example, "$a<_r c$" or "Pacr". *Mutatis mutandis* if parthood$_m$ has four or more argument places.

(ii) Its second premise is a true atomic sentence to the effect that c is a part$_m$ of d, at whatever region or regions (if any) participate in the relevant instance of parthood$_m$. Thus, if parthood$_m$ is two-place, then the second premise of the argument might be the sentence "$c<d$". Etc.

(iii) Its third premise is a true sentence that expresses the transitivity of parthood$_m$ or (if parthood$_m$ is not two-place) the most straightforward and natural analogue of that principle that can be framed in terms of a predicate whose adicity matches that of parthood$_m$. Thus, if parthood$_m$ is two-place, then the third premise might be the sentence "$\forall x\forall y\forall z[(x<y \;\&\; y<z) \rightarrow x<z]$," and if parthood$_m$ is three-place, then the third premise might be "$\forall x\forall y\forall z\forall w[(x<_w y \;\&\; y<_w z) \rightarrow x<_w z]$." *Mutatis mutandis* if parthood$_m$ has four or more argument places.

(iv) Its conclusion is an atomic sentence to the effect that a is a part$_m$ of d, at whatever region or regions (if any) participate in the relevant instance of parthood$_m$. Thus if parthood$_m$ is two-place, then the conclusion of the argument might be the sentence "$a<d$", and if parthood$_m$ is three-place, the conclusion might be the sentence "$a<_{r*}d$", where "r*" is a name for the region (whatever it is) at which a is a part$_m$ of d. *Mutatis mutandis* if parthood$_m$ has four or more argument places.

In sum, the relevant 'transitivity' principle should have a certain amount of power in this case; it should *apply* to the objects described above in such a way as to permit a sound argument of the given type. I take this to be a piece of intuitive data that our theories about parthood$_m$ ought to accommodate. Call arguments of the given type *transitivity arguments*, and call the view that there is such an argument the *transitivity intuition.*

The transitivity intuition is easily accommodated by Absolutism. According to Absolutism, we can formulate a transitivity argument as follows:

Argument 1
Premise 1.1 $a<c$
Premise 1.2 $c<d$
Premise 1.3 $\forall x\forall y\forall z[(x<y \;\&\; y<z) \rightarrow x<z]$
Conclusion 1 $a<d$

The transitivity intuition can also be accommodated by 3P, provided that we do not combine this view with the wrong answer to the Restriction Question. Suppose, for example, that both 3P and the Spacetime Principle are true. Then the following counts as a transitivity argument:

Argument 2
Premise 2.1 $a<_{Rmax}c$
Premise 2.2 $c<_{Rmax}d$
Premise 2.3 $\forall x\forall y\forall z\forall w[(x<_{w}y \;\&\; y<_{w}z) \rightarrow x<_{w}z]$
Conclusion 2 $a<_{Rmax}d$

Similarly, if 3P and the Time-Slice Principle were true, we could formulate a transitivity argument just by replacing each occurrence of "Rmax" above with an occurrence of "Rslice". So, despite their other faults, the Spacetime Principle and the Time-Slice Principle both fare well with respect to the transitivity intuition.

Not so for either the Whole Location Principle or the Part Location Principle. Given 3P and the Whole Location Principle, together with the demand to formulate premises and a conclusion that fit the specifications set out in (i) – (iv), the best we can do is:

Argument 3
Premise 3.1 $a<_{Rc}c$
Premise 3.2 $c<_{Rd}d$
Premise 3.3 $\forall x\forall y\forall z\forall w[(x<_{w}y \;\&\; y<_{w}z) \rightarrow x<_{w}z]$
Conclusion 3 $a<_{Rd}d$

The obvious problem with Argument 3 is that it is not formally valid: the subscript in Premise 3.1 does not match the subscript in Premise 3.2, which it would need to in order for Premise 3.3 (a.k.a. Transitivity3P) to get any traction. But of course we cannot change the argument to fix this without defying the Whole Location Principle. Given the set-up of our case, that principle tells us that the only region at which a is a $part_m$ of c is Rc and that the only region at which c is a $part_m$ of d is a *different* region, Rd. In other words, the Whole Location Principle (unlike the Spacetime Principle or the Time-Slice Principle) tells us that there is no *common* region at which both a is a part$_m$ of c and c is a part$_m$ of d. But without such a common region, we cannot use Transitivity3P to derive a conclusion about a's being a part$_m$ of d.

One might object that our 'transitivity' principle is too weak, and in particular that we should replace it with

Transitivity3P+: $\forall x \forall y \forall z \forall w \forall w^*[(x <_w y \ \& \ y <_{w^*} z) \rightarrow x <_{w^*} z]$

This would make Argument 3 formally valid, but Transitivity3P+ is much too strong to be plausible to any multi-locationist who takes material objects to vary mereologically from one location to another. (Hence feature (iii), which requires that the principle in question be *true*, is not plausibly satisfied.) Presumably any such philosopher will want to allow for the following type of case:

Case 2. Bob is multi-located; he exactly occupies the spacetime regions R1 and R2. As he is at R1, he has both a left hand (Hand) and a left thumb (Thumb) as parts. Hand exactly occupies a proper subregion of R1, namely R1Hand, and Thumb exactly occupies a proper subregion of R1Hand, namely R1Thumb. However, as Bob is at R2, he has Hand as a part (it exactly occupies R2Hand, a proper subregion of R2) but he does not have Thumb as a part. Thumb does not exactly occupy any subregion of R2.

As applied to Case 2, 3P and the Whole Location Principle tell us that Thumb is a part$_m$ of Hand at R1Hand and that Hand is a part$_m$ of Bob at R1 *and* at R2. Together with Transitivity3P+, this entails that Thumb is a part$_m$ of Bob at R2. But this is clearly wrong: Thumb is not a part$_m$ of Bob at R2. Transitivity3P+ is a non-starter.[34]

---

34 One might be tempted to replace Transitivity3P+ with:
Transitivity3P∃: $\forall x \forall y \forall z \forall w \forall w^*[(x <_w y \ \& \ y <_{w^*} z) \rightarrow \exists w^{**}(x <_{w^{**}} z)]$
But this is vulnerable to a similar counterexample. Mary donates a kidney (Kidney3) to John. There is a cell (Cell3) which, intuitively, is a part of Kidney3 only when that kidney is a part of Mary: the cell dies and is shed well before the kidney is donated to John. In that case we might have the following: Kidney3 exactly occupies many regions, among them Rmk, which does not overlap any region that John exactly occupies, and Rjk, which is a proper subregion of a region (Rj) that John exactly occupies. Cell3 exactly occupies many regions, among them Rcell, which is a proper subregion of Rmk, but Cell3 does not exactly occupy any region that overlaps a region that John exactly occupies. Then, given the Whole Location Principle, it is plausible that: (i) Cell3 is a part$_m$ of Kidney3 at Rmk and (ii) Kidney3 is a part$_m$ of John at Rj, and yet, contrary to Transitivity 3P∃, there is *nothing* at which Cell3 is a part$_m$ of John.

A second (and perhaps more obvious) problem for this suggestion is that, even if the principle were plausible (which it is not), it's not strong enough to combine with the sorts of premises I've described to yield, as a conclusion, an atomic sentence like "$a <_{Rd} d$", which

A more promising suggestion is to replace Transitivity3P with a principle somewhat weaker than Transitivity3P+. If we symbolize the predicate 'x is a subregion of y' as '$x \sqsubseteq y$' then we can state the new principle as:

Transitivity3Psub: $\forall x \forall y \forall z \forall w \forall w^*[(x <_{w} y \ \& \ y <_{w^*} z \ \& \ w \sqsubseteq w^*) \rightarrow x <_{w^*} z]$

This principle does not yield the absurd conclusion that Thumb is a part$_{m}$ of Bob at R2. To get that result from Transitivity3Psub, we would also need the claim that R1Hand (the region at which Thumb is a part$_{m}$ of Hand) is a subregion of R2 (this being a region at which Hand is a part$_{m}$ of Bob). But given the set-up of Case 2, this claim is false: recall that Thumb exactly occupies a subregion of R1Hand but does not exactly occupy any subregion of R2. This entails that R1Hand is not a subregion of R2. So Case 2 poses no problem for Transitivity3Psub.

Another point in favor of Transitivity3Psub is that, as applied to Case 1 (involving a, b, c, and d), it *can* be used to show that a is a part$_{m}$ of d at Rd. For consider:

Argument 4	
Premise 4.1	$a <_{Rc} c$
Premise 4.2	$c <_{Rd} d$
Premise 4.3	$\forall x \forall y \forall z \forall w \forall w^*[(x <_{w} y \ \& \ y <_{w^*} z \ \& \ w \sqsubseteq w^*) \rightarrow x <_{w^*} z]$
Premise 4.4	$Rc \sqsubseteq Rd$
Conclusion 4	$a <_{Rd} d$

Argument 4 is formally valid and, given 3P and the Whole Location Principle, all of its premises are overwhelmingly plausible. It may seem, therefore, that Transitivity3Psub allows the defender of the Whole Location Principle to accommodate the transitivity intuition.

But this would be a mistake, for two reasons. First, recall the content of the transitivity intuition – viz., that there is some formally valid, three-premise argument with features (i) – (iv). Argument 4 is no such argument. It has *four* premises, and it obviously cannot be converted into a formally valid, three-premise argument without strengthening one of its first three premises, or weakening its conclusion, in a way that would make it forfeit one of features (i) – (iv).

Second, even as it stands, Argument 4 clearly does not possess feature (iii). I take it that no one would be tempted to suggest that Transitivity3Psub is the most *natural* and *straightforward* three-place counterpart of

---

is what we need. All it would give us is: "$\exists x(a <_{x} d)$". Thanks to Adam Sennet for discussion.

Transitivity2P. Far from it: Transitivity3Psub is cooked up and cobbled together in an attempt to reconcile opposing motivations – the (to my mind well-justified) desire to accommodate the transitivity intuition, and the (to my mind misguided) desire to preserve the Whole Location Principle. The most egregious symptom of this is the fact that the sentence expressing Transitivity3Psub contains an additional predicate, '$x \sqsubseteq y$', over and above '$x <_z y$'. Even if the former ends up being definable in terms of the latter (a related question is addressed at the end of section 5), the fact that the extra material is needed points to the complexity and unnaturalness of the principle.

Admittedly, Argument 4 is much better than nothing. It is valid, it has plausible premises, and it *resembles* a transitivity argument; it comes much closer to satisfying the transitivity intuition than does, say, the invalid Argument 3. So some friends of 3P and the Whole Location Principle may be willing to bite this bullet and deny the letter of the transitivity intuition, given that they have Argument 4 to fall back on. The crucial point is merely that *there is a bullet to bite here.* After all, Absolutists have no trouble satisfying the *letter* of the transitivity intuition. Likewise for friends of 3P who endorse either the Spacetime Principle or the Time-Slice Principle. And likewise for friends of 4P, as I explain in section 4. So I think that I am well within my rights to insist that, other things being equal, a theory of parthood$_m$ that accommodates the transitivity intuition is more plausible than one that does not.

This completes my discussion of the transitivity intuition and its bearing on the Whole Location Principle. I assume that the friend of that principle cannot do any better than Argument 4 with regard to satisfying the transitivity intuition, and that Argument 4 leaves much to be desired.

I will now turn briefly to the package consisting of the Ra view and the Part Location Principle. This package is vulnerable to a parallel argument. For if we accept this package, and if we strictly comply with the demand to formulate premises and a conclusion that fit the specifications set out in (i) – (iv), then our best attempt at a transitivity argument will be:

Argument 5

Premise 5.1	$a <_{Ra} c$
Premise 5.2	$c <_{Rc} d$
Premise 5.3	$\forall x \forall y \forall z \forall w[(x <_w y \ \& \ y <_w z) \rightarrow x <_w z]$
Conclusion 5	$a <_{Ra} d$

This argument, like Argument 3, is not formally valid: the subscripts in the first two premises don't match. Like the Whole Location Principle, the Part Location Principle tells us that there is no common region at which both a is a $part_m$ of c and c is a $part_m$ of d. We can, as before, formulate something that resembles a transitivity argument: we can replace Premise 5.3 with

Premise 5.3*: $\forall x \forall y \forall z \forall w \forall w^*[(x <_w y \;\&\; y <_{w^*} z \;\&\; w \sqsubseteq w^*) \rightarrow x <_w z]$

and add a premise to the effect that Ra is a subregion of Rc. But, like Argument 4, the result will still fall short of being a genuine transitivity argument.[35] So the Part Location Principle fares no better than the Whole Location Principle. Both conflict with the transitivity intuition. This is their main defect.[36]

---

35 We might try strengthening the relevant transitivity principle in a way that is tailored, not to the Whole Location Principle (as was Transitivity3P+), but rather to the Part Location Principle – viz.:

Transitivity3P+**: $\forall x \forall y \forall z \forall w \forall w^*[(x <_w y \;\&\; y <_{w^*} z) \rightarrow x <_w z]$

But this principle is vulnerable to the same counterexample (Case 2) as was Transitivity3P+.

36 A variant of this problem also applies to an appealing pluralist answer to our question about a and c (inspired by Hudson 2001: 61-71). Suppose that one thinks that a is a $part_m$ of c at every region that is both a superregion of a's location and a subregion of c's location. (This includes, among other regions, Ra, Ra+, and Rc.) This view is most plausible in the context of the following general principle:

Tweenism: Necessarily, for any material objects x and y and any spacetime region R, if x is a part of y at R, then R is a superregion of some location of x and a subregion of some location of y.

The Tweenist can plausibly regard the following as a transitivity argument: Premise T1: $a <_{Rc} c$. Premise T2: $c <_{Rc} d$. Premise T3: $\forall x \forall y \forall z \forall w[(x <_w y \;\&\; y <_w z) \rightarrow x <_w z]$. Conclusion: $a <_{Rc} d$. The Tweenist will see T1 as being true on the grounds that Rc is a (proper) superregion of a's location and an (improper) subregion of c's location, and he will see T2 as being true on the grounds that Rc is an (improper) superregion of c's location and a (proper) subregion of d's location. Finally, he will see the conclusion as being true on the grounds that Rc is a (proper) superregion of a's location and a (proper) subregion of d's location. So it appears that Tweenism can accommodate the transitivity intuition as stated.

Suppose, however, that we introduce a further object, e*, that we can think of as being composed of d together with another small object b**, where b**'s lone location, Rb**, is disjoint from Rd, and where e*'s lone location, Re*, is the sum or union of Rb** and Rd. Then it seems that we ought to be able to construct an argument whose conclusion is an atomic sentence to the effect that a is a $part_m$ of e* (at certain region(s), perhaps), and that we ought to be able to construct this argument merely by adding *just one additional premise* to those from our earlier transitivity argument – viz., an atomic sentence to the effect that d is a $part_m$ of e* (at certain region(s), perhaps). The Tweenist cannot satisfy this demand. According to him, the only regions at which d is a $part_m$ of e* are *super*regions of Rd, and none of these is a region at which a is a $part_m$ of c (all of which are *sub*regions of Rc). I.e., on his view, there is no single region at which (i) a is a $part_m$ of c, (ii) c is a $part_m$ of d, and (iii) d is a $part_m$ of e*, although there are regions at which the first two clauses hold and other regions at which the second two clauses hold. As I indicate in note 41, this case poses no problem for 4P.

3.2.4 The Two Case Intuition

There is also a second objection to these two principles that deserves a brief mention. In addition to conflicting with the transitivity intuition, they both conflict with what I will call the *two case intuition*. Consider the situations represented in Fig. 4.[37]

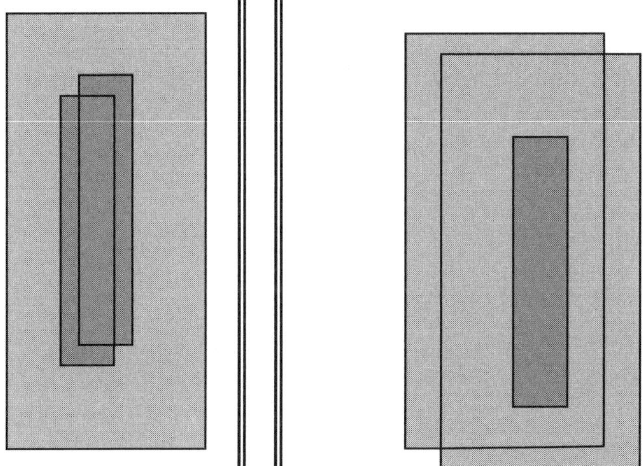

Case 3. Fred is mono-located; he exactly occupies Rf only. Fredpart is bi-located; its only locations are the distinct but overlapping regions Rfsub1 and Rfsub2, each a proper subregion of Rf.

Case 4. Greta is bi-located; she exactly occupies the distinct but overlapping regions Rg1 and Rg2 only. Gretapart is mono-located; its only location is Rgsub, a proper subregion of Rg1 and of Rg2.

Figure 4

37 According to most multi-locationists (endurantists and MROists), these situations will be at best extremely remote possibilities, and so the failure to accommodate them will hardly seem to be much of a vice. The one multi-locationist who would seem to be committed to their possibility is Hudson (2001). On his view, as I noted earlier, ordinary objects are temporally extended 'worms' that typically have many (mostly overlapping) 4D locations. The idea, roughly, is to say that each of the many precisely demarcated regions that we would normally regard as a *candidate* for being my (e.g.) location actually *is* one of my locations. To allow for Case 3 below, in which the larger object Fred is mono-located (while Fredpart is multi-located), we would need to suppose that for some reason, the usual sources of boundary vagueness are absent for Fred but are present for Fredpart. (Perhaps Fred's surface is especially tightly integrated and sharply demarcated from its surroundings.) Case 4 is easier. Just let the smaller object Gretapart be a mono-located simple, and let Greta be multi-located in the way Hudson regards as typical.

In Case 3, there seem to be exactly two instances of parthood_m involving Fredpart and Fred: the former object seems to be part_m of the latter object twice over. The Part Location Principle allows for this. Since Fredpart has two different locations, the Part Location Principle lets us say that Fredpart is part_m of Fred both at Rfsub1 and at Rfsub2. In other words, it lets us say that two different ordered triples involving Fredpart and Fred (⟨Fredpart, Fred, Rfsub1⟩ and ⟨Fredpart, Fred, Rfsub2⟩) are such that their members instantiate parthood_m in the order specified by the triple. The Whole Location Principle, however, has trouble with Case 3. Since Fred has only one location, this principle tells us that there is at most one region at which Fredpart is a part_m of Fred, hence that there is at most one instance of parthood_m associated with the given pair of objects.

Case 4 elicits similar intuitive judgments: there seem to be two instances of parthood_m involving Gretapart and Greta. But in this case, it is the Whole Location Principle that allows for the intuitively correct verdict and the Part Location Principle that fails. Since Greta exactly occupies both Rg1 and Rg2, the Whole Location Principle lets us say: Gretapart is a part_m of Greta both at Rg1 and at Rg2. On the other hand, since Gretapart has only one location, the Part Location Principle forces us to say that there is at most one region at which Gretapart is a part_m of Greta, hence that there is at most one instance of parthood_m involving these two objects.

The two case intuition, then, is the following conjunction: there are two instances of parthood_m involving Fredpart and Fred, and there are two instances of parthood_m involving Gretapart and Greta. Neither the Part Location Principle nor the Whole Location Principle can handle both conjuncts of the intuition, although each principle handles one conjunct.

Unlike the previous intuitions that I have discussed, the two case intuition poses problems for virtually all of the views about parthood_m considered so far. According to Absoluteness, there is only one ordered n-tuple involving Fredpart and Fred whose members instantiate parthood_m in the order specified – namely, the ordered pair ⟨Fredpart, Fred⟩. Thus it tells us that there is only one instance of parthood_m involving these objects. *Mutatis mutandis* for Gretapart and Greta. Absoluteness fails for both cases. (Of course, it is highly unlikely that any Absolutist would accept the possibility of Cases 3 or 4, so we should be hesitant to think of these cases as the basis of a non-question-begging *argument* against Absolutism.)

Our 3P-ist views divided into two categories: pluralist and non-pluralist. As for the pluralist answers to our original question about a and c, it seems that on any remotely plausible way of generalizing these answers, we would get the result that Fredpart is a part_m of Fred *more* than twice (and likewise for Gretapart and Greta). But having too many of the relevant instances is no better than having too few.

What about the Spacetime Principle? On the assumption that the regions involved in Cases 3 and 4 all belong to a single spacetime, this principle gives us the result that Fredpart is a $part_m$ of Fred at most once, and likewise for Gretapart and Greta. Similarly, on the assumption that these regions are all subregions of exactly one common global time-slice, the Time-Slice Principle delivers this same result. So the problems posed by the two case intuition are not specific to the Whole Location Principle or the Part Location Principle. And yet we will see that these problems can be avoided if we reject 3P in favor of 4P.

4. Four-Place Parthood

According to 4P, $parthood_m$ is a four-place relation that can be expressed by the predicate 'x at w is a $part_m$ of y at z'. Given multi-location and 4P, one natural response to the Restriction Question[38] is to suggest that $parthood_m$ is governed by

LLP: Necessarily, for any material objects x and y and any spacetime regions w and z, if x at w is a $part_m$ of y at z, then x exactly occupies w and y exactly occupies z.[39]

I will now argue that LLP avoids all of the problems discussed in section 3. In particular, it respects (i) the single case intuition, (ii) the intrinsicness intuition, (iii) the transitivity intuition, and (iv) the two case intuition.

We can start with (i). As applied to Case 1, LLP tells us that there is just one ordered n-tuple whose members instantiate $parthood_m$ (in the order given by the n-tuple) – namely, the ordered quadruple ⟨a, Ra, c, Rc⟩. After all, a exactly occupies only Ra, and c exactly occupies only Rc. So, with just one such n-tuple, we get the result that there is just one instance of $parthood_m$ involving a and c; in slightly different terms, a is a $part_m$ of c just once. Moreover, the suggestion that ⟨a, Ra, c, Rc⟩ is the one and only n-tuple with the relevant feature cannot be accused of being arbitrary or of making $parthood_m$ seem haphazard. Ra and Rc are quite special as far as a and c are concerned, for obvious reasons.

38 Or a variant of that question framed in terms that are more appropriate to 4P.

39 We might also add " . . . and w is a subregion of z." I suspect that anyone who finds LLP plausible will find the strengthened version roughly equally plausible.

Next consider (ii), the intrinsicness intuition. Here again, LLP fares well. As I noted earlier, being a spacetime region is plausibly an intrinsic property and exact occupation is plausibly an intrinsic relation. So it would seem that if one ordered quadruple of entities is eligible for parthood_m according to LLP, then any plurally duplicate quadruple is also eligible. In other words, LLP never treats plurally duplicate ordered quadruples differently. So far as LLP is concerned, then, its proponents are free to hold that parthood_m is an intrinsic relation.

We can turn now to (iii), the transitivity intuition. It seems to me that given 4P and LLP, one principle stands out as the most straightforward and natural analogue of Transitivity2P. If we symbolize the predicate 'x at w is a part_m of y at z' as '$x_w <_z y$', then we can formulate that principle as

Transitivity4P: $\forall x \forall r1 \forall y \forall r2 \forall z \forall r3[(x_{r1} <_{r2} y \;\&\; y_{r2} <_{r3} z) \rightarrow x_{r1} <_{r3} z]$

In words, this says that if x at r1 is a part_m of y at r2 and if y at r2 is a part_m of z at r3, then x at r1 is a part_m of z at r3.[40] To make the analogy easier to grasp, it may be helpful to define a two-place relation of *pair-parthood*:

D5: p is a pair-part of p* =df. $\exists x \exists r1 \exists y \exists r2[p=\langle x, r1\rangle \;\&\; p^*=\langle y, r2\rangle \;\&\; x_{r1} <_{r2} y]$.

Then, given the existence of the relevant pairs, Transitivity4P is equivalent to the claim that pair-parthood is transitive in the strict sense.

Suppose, then, that Transitivity4P is a sufficiently natural analogue of Transitivity2P. In that case, the friend of 4P will say that a transitivity argument can be formulated as follows:

Argument 6
Premise 6.1 $a_{Ra} <_{Rc} c$
Premise 6.2 $c_{Rc} <_{Rd} d$
Premise 6.3 $\forall x \forall r1 \forall y \forall r2 \forall z \forall r3[(x_{r1} <_{r2} y \;\&\; y_{r2} <_{r3} z) \rightarrow x_{r1} <_{r3} z]$

40 Obviously, there are many vaguely transitivity-like principles that can be framed in terms of our four-place parthood predicate. I suspect that the most serious rival to Transitivity4P is Transitivity4P*: $\forall x \forall y \forall z \forall r[(x_r <_r y \;\&\; y_r <_r z) \rightarrow x_r <_r z]$
Transitivity4P* says that if x at r is a part_m of y at r, and y at r is a part_m of z at r, then x at r is a part_m of z at r. This is quite weak. If x and y are material objects and r is a spacetime region, then, given LLP, x at r is a part_m of y at r only if x and y both exactly occupy r. So a *trio* of material objects can satisfy the antecedent of Transitivity4P* only if they all share a common location – i.e., only if they all exactly occupy the very same region. It would come as a big surprise, I think, if this were the strongest transitivity-like principle that parthood_m obeys. Typically we think that this relation is governed by a transitivity-like principle that applies to trios of *non*-co-located objects.

Conclusion 6 $a_{Ra} <_{Rd} d$

Premise 6.1 says that a at Ra is a part$_m$ of c at Rc, and Premise 6.2 says that c at Rc is a part$_m$ of d at Rd. Since Ra, Rc, and Rd are locations of a, c, and d, respectively, friends of LLP will find these premises highly plausible. Together with Premise 6.3, they yield a formally valid argument for Conclusion 6, which says that a at Ra is a part$_m$ of d at Rd. This is exactly the sort of conclusion concerning a and d that we initially hoped to be able to prove. It should be clear, therefore, that if we adopt 4P and LLP, we can easily accommodate the transitivity intuition.[41]

Finally, consider (iv), the two case intuition. It says that there are exactly two instances of parthood$_m$ involving Fredpart and Fred (in Case 3) and exactly two instances of parthood$_m$ involving Gretapart and Greta (in Case 4). Unlike any of the other views that we have considered so far, the 4P+LLP package allows for this without any strain at all.

In Case 3, Fredpart has exactly two locations, Rfsub1 and Rfsub2, and Fred has exactly one location, Rf. Together with LLP, this allows us to say that

- Fredpart at Rfsub1 is a part$_m$ of Fred at Rf, and
- Fredpart at Rfsub2 is a part$_m$ of Fred at Rf;

and it entails that (i) for no x aside from Rfsub1 or Rfsub2 is there a y such that Fredpart at x is a part$_m$ of Fred at y and (ii) for no y aside from Rf is there an x such that Fredpart at x is a part$_m$ of Fred at y. In other words, there are no ordered n-tuples involving Fredpart and Fred aside from ⟨Fredpart, Rfsub1, Fred, Rf⟩ and ⟨Fredpart, Rfsub2, Fred, Rf⟩ whose members instantiate parthood$_m$ in the order given by the n-tuple.[42] On the assumption that there are exactly two such n-tuples, we get the result that there are exactly two instances of parthood$_m$ involving Fredpart and Fred – i.e., that Fredpart is a part$_m$ of Fred exactly twice. And of course this is what we wanted.

Parallel remarks apply to Case 4, in which Gretapart has exactly one location, Rgsub, and Greta has exactly two locations, Rg1 and Rg2. Together with LLP, this lets us say that

- Gretapart at Rgsub is a part$_m$ of Greta at Rg1, and
- Gretapart at Rgsub is a part$_m$ of Greta at Rg2;

---

41 Likewise, the augmented case described in note 36 poses no problem. We simply add the following premise: $d_{Rd} <_{Re*} e*$. The result is a formally valid four-premise argument for the conclusion, "$a_{Ra} <_{Re*} e*$", as desired.

42 More carefully: given 4P and LLP, there are no such ordered n-tuples *in which Fredpart is the first element and Fred is the third element* aside from the given pair of ordered quadruples. LLP is silent, for example, as to whether (*) Rfsub1 at Fredpart is a part$_m$ of Rf at Fred. But I take it that even if (*) is true, we still have the result that Fredpart is a part$_m$ of Fred exactly twice.

and it entails that there are no ordered n-tuples involving Gretapart and Greta aside from ⟨Gretapart, Rgsub, Greta, Rg1⟩ and ⟨Gretapart, Rgsub, Greta, Rg2⟩ whose members instantiate parthood$_m$ in the order given by the n-tuple.[43] Thus we are free to say that there are exactly two instances of parthood$_m$ involving Gretapart and Greta – i.e., that Gretapart is a part$_m$ of Greta exactly twice. 4P+LLP has no trouble handling the two case intuition.

Let me summarize my results so far. Given 3P, every remotely plausible answer to the Restriction Question violates *both* the two case intuition *and* (more importantly) either the single case intuition, the intrinsicness intuition, or the transitivity intuition. Given 4P, however, there is at least one quite natural answer to the Restriction Question that respects all four of those intuitions. This gives the multi-locationist a very powerful reason to prefer 4P to 3P.

Are there any reasons to have the opposing preference? In my view, the most serious objection to 4P is this. Other things being equal, the more closely a theory of parthood$_m$ resembles Absolutism, the better. (This principle was invoked in support of the single case intuition and the intrinsicness intuition.) The most obvious respect in which a theory of parthood$_m$ can resemble Absolutism is with regard to the adicity that the theory attributes to parthood$_m$. But in that respect, 3P clearly resembles Absolutism more closely than does 4P: after all, 3 is closer to 2 than 4 is!

I concede that, *ceteris paribus*, we should try to minimize the number of extra argument places that we posit in parthood$_m$ (beyond the ordinary two). But, first, it seems to me that this consideration is rather weak and easily overridden by the other factors that I have discussed.

Second, 4P may resemble Absolutism more closely than does 3P *even with regard to considerations about argument places themselves*. After all, Absolutism tells us that the argument places in parthood$_m$ are evenly divided between (i) those that are especially closely associated with 'the part' and (ii) those that are especially closely associated with 'the whole.' 4P, when combined with LLP, tells us exactly the same thing. For according to 4P+LLP, parthood$_m$ has four argument places, two of which are reserved for the part or one of its locations, and the other two of which are reserved for the whole or one of *its* locations. Obviously 3P cannot match this. So, even with regard to issues about the argument places themselves, 4P arguably does a better job of mimicking Absoluteness than does 3P.

Finally, we should note that even if parthood$_m$ is a four-place relation, we can use it to define various non-fundamental parthood relations that

43 More carefully: given 4P and LLP, there are no such ordered n-tuples *in which Gretapart is the first element and Greta is the third element* aside from the given pair of ordered quadruples. See the previous note.

have fewer argument places. Here, for example, is a definition of a two-place relation that seems to deserve the title *parthood simpliciter*:

Part *Simpliciter:* x is a part *simpliciter* of y =df. $\exists$r(x exactly occupies r) & $\forall$r[x exactly occupies r $\rightarrow$ $\exists$r2($x_r <_{r2} y$)]

To say that x is a part *simpliciter* of y, according to this definition, is to say that x has a location, and for any such location r, x at r is a part$_m$ of y, at some location of y. In other words, x can be found somewhere, and wherever x can be found, it will there be a part$_m$ of y, at some location of y.

Likewise, we can define various relations that are three-place and that behave in accordance with the principles discussed in section 3. Consider, for example, the following definitions, which I take to require no unpacking:

Time-Slice Parthood: x is a part$_{ts}$ of y at t =df. (i) t is a global time-slice, (ii) there are subregions r1 and r2 of t such that x at r1 is a part$_m$ of y at r2, and (iii) for any subregion r of t, if x exactly occupies r, then there is some subregion r* of t such that x at r is a part$_m$ of y at r*.

Part Location Parthood: x is a part$_{pl}$ of y at r =df. r is a spacetime region, and there is some spacetime region r* such that x at r is a part$_m$ of y at r*.

Given the availability of definitions like these, the friend of 4P need not deny the existence or intelligibility of the relevant two-place or three-place relations; she need only deny their fundamentality.

This completes the case for preferring 4P over 3P, given multi-location.

5. Formal Considerations

Many philosophers believe that if parthood$_m$ is a two-place relation, then it has certain familiar formal properties, such as reflexivity and transitivity. What should these philosophers say about the formal properties of parthood$_m$ on the assumption that it is a four-place relation? In this section I will tentatively recommend certain answers to that question.

I will not attempt to argue for or against any claim of the form 'if parthood$_m$ is two-place, then it has formal property P' or 'if parthood$_m$ is four-place, then it has formal property Q.' For example, I neither endorse nor reject the claim that if parthood$_m$ is two-place, then it obeys a Uniqueness of Composition principle.

Rather, what I will attempt to do is to begin to address questions like the following: "Suppose that one has no settled opinion as to the adicity of parthood$_m$, but one is convinced that *if* parthood$_m$ is two-place, then it obeys a Uniqueness of Composition principle. In that case, is there any analogous principle that one will find it natural to adopt if one becomes convinced that parthood$_m$ is in fact a *four*-place, rather than a two-place, relation? If so, what is the principle?" My ambitions here are modest: I hope only to uncover some considerations that provide guidance on these questions, not to settle them.

I will begin by setting out some familiar definitions and principles framed in terms of a two-place parthood predicate; I will then suggest what I take to be natural analogues of these principles, framed in terms of a four-place parthood predicate. First, some definitions:

Proper Part:	$x \ll y$ =df. $x<y$ & $x \neq y$
Overlap:	xOy =df. $\exists z(z<x$ & $z<y)$
Disjointness:	xDy =df. $\neg xOy$
Fusion:	xFs =df. $\exists y(y \in s)$ & $\forall y[yOx \leftrightarrow \exists z(z \in s$ & $yOz)]$

A proper part of a thing is a part of a thing that is not identical with that thing; things overlap iff they share a common part; and they are disjoint iff they don't overlap. The fusion predicate is here defined in terms of set membership[44] and overlap: x fuses s just in case (i) s has at least one member and (ii) a thing overlaps x just in case it overlaps some member of s.

Next, some principles. Most would agree that if parthood$_m$ is two-place, then it is governed at least by the following:

Reflexivity:	$\forall x(x<x)$
Transitivity:	$\forall x \forall y \forall z[(x<y$ & $y<z) \rightarrow x<z]$
Weak Supplementation (WSP):	$\forall x \forall y[(x \ll y) \rightarrow \exists z(z \ll y$ & $zDx)]$

44 A fusion predicate can also be defined schematically (see Hovda (2009) for discussion) or by appeal to plural quantifiers and variables (x fuses the Ys =df. $\forall z[zOx \leftrightarrow \exists y($y is among the Ys & $yOz)])$. See Lewis (1991: 73) for a different definition of 'fusion' in terms of plural quantifiers and variables.

Of these, I take it that only WSP requires comment: it says that if a thing has a proper part, then it has a second proper part that is disjoint from the first.[45]

Some philosophers think that parthood$_m$, if two-place, is governed not merely by the three principles above but also by one or both of the following:

Universalism: $\forall s[\exists x(x \in s) \rightarrow \exists x(xFs)]$
Uniqueness: $\forall s \forall x \forall y[(xFs \ \& \ yFs) \rightarrow x=y]$

Universalism says that every non-empty set has at least one fusion; Uniqueness says that nothing has more than one fusion.

Can we formulate analogues of these principles that are appropriate to a four-place parthood relation? What further principles, if any, govern the interaction between our four-place parthood relation and the subregion and exact occupation relations?

I have already made forays in this direction. I have suggested that parthood$_m$ is governed by Transitivity4P and that its interaction with exact occupation is plausibly governed by LLP. To state these and other principles formally, I will continue to use '$x_{r1}<_{r2}y$' for 'x at r1 is a part$_m$ of y at r2' and '$r1 \sqsubseteq r2$' for 'r1 is a subregion of r2', and I will introduce two new symbols: 'Rr' for 'r is a spacetime region' and '$x \otimes r$' for 'x exactly occupies r'. This lets us write out a non-modal variant of LLP as:

LLP*: $\forall x \forall r1 \forall y \forall r2[x_{r1}<_{r2}y \rightarrow (x \otimes r1 \ \& \ y \otimes r2)]$

45 This is prominently discussed in Simons (1987); it is meant to be (like Reflexivity and Transitivity) acceptable to friends and foes of Uniqueness and to friends and foes of Universalism. But some foes of Uniqueness (e.g., Thomson 1998) may doubt it for the following reason. Suppose that Lump and Goliath are two different fusions of the same particles. Suppose further that at least one of them, say Lump, is a part of the other, Goliath. (Thomson takes each of them to be a part of the other.) Then, since they are non-identical, Lump is a *proper* part of Goliath. But, contrary to WSP, Goliath does not have any parts that are disjoint from Lump, since they are both fusions of the same particles. There is a variant of WSP, however, that captures the core intuition underlying WSP while avoiding this worry:

Quasi-Supplementation (QS): $\forall x \forall y[(x<y \ \& \ x \neq y) \rightarrow \exists z \exists w(z<y \ \& \ w<y \ \& \ zDw)]$

QS says that if a thing has a part with which it is not numerically identical, then it has parts that are disjoint from *each other*, though not necessarily from the original part. This captures Simons's important insight (used to motivate WSP) that "surely if a universe is complex (i.e. has proper parts at all) then at least two of these parts will be disjoint" (1987: 27). The Thomson-inspired treatment of Lump and Goliath poses no threat to QS: clearly each of those objects has disjoint parts (e.g., a particle in Goliath's left foot and one in his right foot).

This says that if x at r1 is a part$_m$ of y at r2, then x exactly occupies r1 and y exactly occupies r2. I suspect that many philosophers will also find it highly plausible that if x at r1 is a part$_m$ of y at r2, and if r1 and r2 are both spacetime regions,[46] then r1 will be a subregion of r2. We can express this formally as follows:

Inheritance: $\forall x \forall r1 \forall y \forall r2[(x_{r1}<_{r2}y \ \& \ Rr1 \ \& \ Rr2) \rightarrow r1 \sqsubseteq r2]$

Inheritance is meant to capture the thought that parts lie within their wholes. We will have more to say about the interaction of parthood$_m$ and exact occupation below.

Consider next the relatively uncontroversial view that parthood$_m$ is governed by a reflexivity principle or by some close analogue of it. A two-place parthood relation could be reflexive in the strict sense: $\forall x(x<x)$. To formulate a counterpart of this principle appropriate to our four-place relation, we will need to take LLP* into account. Thus we should *not* say: $\forall x \forall r(x_r<_r x)$. Together with LLP*, that would entail that everything exactly occupies everything: $\forall x \forall r(x \otimes r)$! Instead we should say that a thing is a part of itself at each of its locations. In other words, if a thing exactly occupies something, then the thing, there, is a part$_m$ of itself, there:

Reflexivity4P: $\forall x \forall r(x \otimes r \rightarrow x_r<_r x)$

It may be helpful to return to the relation of pair-parthood defined in section 4:

p is a pair-part of p* =df. $\exists x \exists r1 \exists y \exists r2(p=\langle x, r1 \rangle \ \& \ p^*=\langle y, r2 \rangle \ \& \ x_{r1}<_{r2}y)$

Setting aside the possibility that a thing x at a location r1 is a part$_m$ of a thing y at a location r2 even though one or both of the ordered pairs ⟨x, r1⟩ and ⟨y, r2⟩ fails to exist, Reflexivity4P is equivalent to the claim that

46 I include this clause because I do not want to rule out the possibility that there are entities other than spacetime regions that can serve as the relata (in the 'subscripted positions') of parthood$_m$. In my (2007: 191) I suggested that one material object, at a given *moment of its personal time*, can be a part$_m$ of another material object, at a given moment of *its* personal time, where these moments of personal time are presumably not to be identified with spacetime regions. But it seems wrong to suggest that the first such moment must therefore be a subregion of the second such moment. Likewise, one might think that there are such things as (non-spatiotemporal) *argument places* in properties and relations, and perhaps even in the entities (if such there be) expressed by sentential operators; one might think that these argument places, though obviously non-spatiotemporal, are quite literally *exactly occupied* by various entities; and one might think that they too can serve as the relata of parthood$_m$. But again it seems wrong to suggest that such argument places would be subregions of one another.

pair-parthood is reflexive over the domain of *occupation pairs*, where an occupation pair is an ordered pair whose first element exactly occupies its second element.

5.1 Weak Supplementation

So far we have stated counterparts of what I take to be the two most widely accepted mereological principles: Transitivity and Reflexivity. The third relatively uncontroversial principle mentioned above is WSP. To state a counterpart of WSP appropriate to four-place parthood, it will be convenient to have a new technical term. We will say that a two-place relation R is *weakly supplementive* if and only if $\forall x \forall y[(xRy \ \& \ x \neq y) \rightarrow \exists z[(zRy \ \& \ z \neq y) \ \& \ \neg\exists w(wRx \ \& \ wRz)]]$. WSP tells us, in the shorthand permitted by our definitions, that *parthood$_m$* is weakly supplementive.

To determine how to formulate the most natural 4P-appropriate counterpart of WSP, I suggest that we look to the examples already established by Transitivity4P and Reflexivity4P. If we set aside the possibility that a thing x at a location r1 is a part$_m$ of a thing y at a location r2 despite the non-existence of ⟨x, r1⟩ or ⟨y, r2⟩, Transivity4P is equivalent to the claim that pair-parthood is transitive, and Reflexivity4P is equivalent to the claim that pair-parthood is reflexive over the domain of occupation pairs. I suggest, then, that the most natural 4P-appropriate counterpart of WSP will be a principle that is equivalent (setting aside the aforementioned possibility) to the claim that pair-parthood is weakly supplementive, at least over the domain of occupation pairs. The most straightforward way to state such a principle, it seems to me, is as follows:

WSP$_{4P}$: $\forall x \forall r1 \forall y \forall r2[(x_{r1}{<_{r2}}y \ \& \ (x \neq y \vee r1 \neq r2)) \rightarrow \exists z \exists r3[z_{r3}{<_{r2}}y \ \& \ (z \neq y \vee r3 \neq r2) \ \& \ \neg\exists w \exists r4(w_{r4}{<_{r3}}z \ \& \ w_{r4}{<_{r1}}x)]]$

To see that this is equivalent to the claim that pair-parthood is weakly supplementive, consider an arbitrarily chosen foursome of entities, a, ra, b, and rb, and suppose:

(i) $a_{ra}{<_{rb}}b \ \& \ (a \neq b \vee ra \neq rb)$

(i) is equivalent to:

(ii) ⟨a, ra⟩ is a pair-part of ⟨b, rb⟩ & ⟨a, ra⟩ ≠ ⟨b, rb⟩

WSP$_{4P}$, together with (i), entails

(iii) $\exists z \exists r3[z_{r3} <_{rb} b \; \& \; (z \neq b \lor r3 \neq rb) \; \& \; \neg \exists w \exists r4(w_{r4} <_{r3} z \; \& \; w_{r4} <_{ra} a)]$

And the claim that pair-parthood is weakly supplementive, together with (ii), entails

(iv) $\exists$p[p is a pair-part of ⟨b, rb⟩ & p≠⟨b, rb⟩ & ¬$\exists$p*(p* is a pair-part of p & p* is a pair-part of ⟨a, ra⟩)]

But (iii) and (iv) are equivalent! So WSP_{4P} can be derived from the claim that pair-parthood is weakly supplementive, and (setting aside the possibility of missing ordered pairs) vice versa. Again, I take this to be good evidence that WSP_{4P} is the closest 4P-appropriate counterpart of Weak Supplementation.

We can state this principle in a more compact form if we define four-place counterparts of the familiar two-place predicates for proper parthood, overlap, and disjointness:

Proper $Part_{4P}$: $x_{r1} \ll_{r2} y$ =df. $x_{r1} <_{r2} y \; \& \; (x \neq y \lor r1 \neq r2)$

$Overlap_{4P:}$ $x_{r1} O_{r2} y$ =df. $\exists z \exists r3(z_{r3} <_{r1} x \; \& \; z_{r3} <_{r2} y)$

$Disjointness_{4P:}$ $x_{r1} D_{r2} y$ =df. $\neg x_{r1} O_{r2} y$

This lets us abbreviate WSP_{4P} as

WSP_{4Pa}: $\forall x \forall r1 \forall y \forall r2[x_{r1} \ll_{r2} y \rightarrow \exists z \exists r3(z_{r3} \ll_{r2} y \; \& \; z_{r3} D_{r1} x)]$[47]

Before we move on, it may be worthwhile to note a potential application of the suggestion that WSP_{4P} is the closest 4P-appropriate counterpart of Weak Supplementation.

In a recent paper, Nikk Effingham and John Robson (2007) argue that endurantism is threatened by the following case:

A certain brick, *Brick₁*, travels backward in time repeatedly, so that it exists at a certain time, t_{100}, 'many times over'. At that time there exist what appear to be one hundred bricks, call them *Brick*$_1$ . . . *Brick*$_{100}$, though in fact each of them is identical to $Brick_1$ (on one or another of its journeys to

47 Similarly, the closest 4P-appropriate counterpart of QS (discussed in note 45) is:
QS_{4P}: $\forall x \forall r1[\exists y \exists r2(y_{r2} <_{r1} x \; \& \; (x \neq y \lor r1 \neq r2)) \rightarrow \exists y \exists r2 \exists z \exists r3(y_{r2} <_{r1} x \; \& \; z_{r3} <_{r1} x \; \& \; y_{r2} D_{r3} z)]$.

the time t_{100}), and a bricklayer arranges 'them' into what appears to be a brick wall, *Wall.*

Effingham and Robson write that

> There is a principle of mereology known as the Weak Supplementation Principle (WSP) which states that every object with a proper part has another proper part that does not overlap the first. If $Brick_1$, $Brick_2$, . . ., $Brick_{100}$ composed a wall, WSP would be false. Consider: any object that was a part of the wall would have to overlap some brick, and as every brick is $Brick_1$ if that object overlaps some brick it overlaps $Brick_1$. Therefore if at t_{100} $Brick_1$, $Brick_2$, . . ., $Brick_{100}$ composed a wall, there would be no object that could be a proper part of the wall that does not overlap $Brick_1$. Given $Brick_1$ is a proper part of that wall, WSP would then be false (2007: 634-635).

If what I have suggested so far is correct, then the endurantist *who accepts multi-location*[48] has a clear response. He should begin by noting that he (qua multi-locationist) has powerful reasons for holding that $parthood_m$ is a four-place relation, where these reasons are completely independent of worries about time travel scenarios. He should then note that the core intuition underlying WSP will give rise to different principles, depending upon one's view about the adicity of $parthood_m$: for example, in conjunction with the view that $parthood_m$ is two-place, that intuition points toward WSP itself. But in conjunction with the view that $parthood_m$ is four-place, that intuition points toward WSP_{4Pa}. Finally, he should note that Effingham and Robson's time travel case poses no threat at all to WSP_{4Pa}. Their case respects that principle.

To see this, recall that instants of time, for the multi-locationist, are merely spacetime regions of a special sort – global time slices. So t_{100} will be such a region. Presumably Wall exactly occupies a certain subregion of t_{100} – call it *Rw* – and $Brick_1$ exactly occupies precisely one hundred different non-overlapping, brick-shaped subregions of Rw – call them R_1. . . R_{100} – where Rw is the sum or union of these regions. In that case, we can say that $Brick_1$ at R_1 is not merely a *part$_m$* of Wall at Rw, but further, a *proper* $part_m$, since it's true that either $Brick_1 \neq Wall$ or $R1 \neq Rw$. (In fact both non-identities hold.) So we can write:

(a) $Brick1\ {}_{R1}{\ll}_{Rw}\ Wall$

48 As we saw in section 3, if an endurantist rejects multi-location and takes instants of time to be simple, *sui generis* entities, rather than spacetime regions of a certain sort, then 3P remains tenable for him. As a result, he apparently has no *independent* motivation to eschew WSP (or its 3P-appropriate counterpart) in favor of WSP_{4Pa}. He may, of course, appeal to the Effingham-Robson case, together with plausibility of endurantism, as motivation enough for shifting to 4P and its associated principles.

The conjunction of (a) and WSP_{4Pa} entails that some z, at some r3, is also a proper part of Wall at Rw and is disjoint from $Brick_1$ at R_1:

(b) $\exists z \exists r3(z_{r3} \ll_{Rw} Wall \ \& \ z_{r3} D_{R1} Brick_1)$

It may initially seem that (b) is false. After all, aren't Effingham and Robson right when they say that every part of Wall overlaps $Brick_1$? Presumably there is a sense in which this is right, but that doesn't show that (b) is false. Given our definitions, (b) tells us that there is a z and an r3 of which the following hold:

(b_1) $z_{r3} <_{Rw} Wall \ \& \ (z \neq Wall \lor r3 \neq Rw)$
(b_2) $\neg \exists w \exists r4(w_{r4} <_{r3} z \ \& \ w_{r4} <_{R1} Brick_1)$

To see there are such things, begin with (b_1). Can we find a z and an r3 such that z at r3 is a $part_m$ of Wall at Rw but either z≠Wall or r3≠Rw? That's easy. Pick anything that's intuitively a proper part of Wall at t_{100} and any subregion of Rw that the thing exactly occupies. Here are two ordered pairs that do the job:

First pair: ⟨G, R_{1subG}⟩, where G is a particular grain of sand that helps to compose $Brick_1$ throughout its career, and R_{1subG} is a tiny, grain-of-sand-shaped subregion of the brick-shaped region R_1, and G exactly occupies R_{1subG}.

Second pair: ⟨$Brick_1$, R_{100}⟩, where R_{100} is a brick-shaped region on the opposite end of Wall from R_1, and R_{100} is exactly occupied by $Brick_1$.

Since G at R_{1subG} is a $part_m$ of Wall at Rw and either G≠Wall or $R_{1subG} \neq Rw$ (in fact both non-identities hold), the first pair does the job. Likewise for the second pair.

But does either of these pairs satisfy (b_2)? Begin with the first pair. Here we need to ask: is there a w and an r4 such that w at r4 is a $part_m$ of G at R_{1subG} and w at r4 is a $part_m$ of $Brick_1$ at R_1? If the answer is "Yes", then the first pair does not satisfy (b_2); otherwise it does. The answer is "Yes". Here is such a w and an r4: G and R_{1subG}. After all, G at R_{1subG} is a $part_m$ of G at R_{1subG} (by Reflexivity4P) and G at R_{1subG} is a $part_m$ of Brick at R_1 (which I take it we can see intuitively, given that R_{1subG} is a subregion of R_1). In other words, G at R_{1subG} is not *disjoint* from $Brick_1$ at R_1.

So let us try the second pair. We must ask: is there a w and an r4 such that w at r4 is a $part_m$ of $Brick_1$ at R_{100} and w at r4 is a $part_m$ of $Brick_1$ at

R_1? No. Consider all the ordered ⟨w, r4⟩ pairs such that w at r4 is a $part_m$ of Brick *at R_1*. On the assumption that any such r4 is a spacetime region, Inheritance tells us that any such r4 must be a *subregion* of R_1. Now consider all the ordered ⟨w, r4⟩ pairs such that w at r4 is a $part_m$ of Brick *at R_{100}*. On the assumption that any such r4 is a spacetime region, Inheritance tells us that any such r4 must be a subregion *of R_{100}*. So, in order for it to be true that there is a w and an r4 such that w at r4 is a $part_m$ of Brick both at R_1 and at R_{100}, those last two regions would need to have a common subregion. But since R_1 and R_{100} are on opposite ends of the wall, they have no common subregion. So we can conclude that there is no pair ⟨w, r4⟩ such that w at r4 is a $part_m$ of $Brick_1$ *both* at R_1 and at R_{100}. In other words, $Brick_1$ at R_{100} *is* disjoint from $Brick_1$ at R_1, in the sense of 'disjoint' relevant to WSP_{4Pa}, and defined above. Thus (b) is true, and the Effingham-Robson case ends up respecting WSP_{4Pa}. (It also respects the weaker principle QS_{4P}, mentioned in note 47.)

Four-place parthood, here argued for on independent grounds, has a beneficial by-product: it defuses what initially appeared to be a serious threat to endurantism. This completes my discussion of Weak Supplementation and its 4P-appropriate counterpart.

5.2 Uniqueness

Now we can turn to the task of stating 4P-appropriate counterparts of the much more controversial Universalism and Uniqueness principles. Both principles will employ a three-place fusion predicate that can be defined as follows:

$Fusion_{4P}$: $x_{r1}Fs =df. \exists y(y\in s) \; \& \; \forall y[y\in s \rightarrow \exists z\exists r2(y=\langle z, r2\rangle)] \; \&$
$\forall y\forall r2[y_{r2}O_{r1}x \leftrightarrow \exists z(z\in s \; \& \; \exists w\exists r3(z=\langle w, r3\rangle \; \& \; y_{r2}O_{r3}w))]$

According to this definition, to say that a thing x fuses a set s at a region r1 is to say that: (i) s is a non-empty set of ordered pairs, and (ii) for any y and any r2, y at r2 overlaps x at r1 if and only if there is a w and an r3 such that: y at r2 overlaps w at r3, and ⟨w, r3⟩ is a member of s. This lets us state a counterpart of Uniqueness:

$Uniqueness_{4P}$: $\forall s\forall x\forall r1\forall y\forall r2[(x_{r1}Fs \; \& \; y_{r2}Fs) \rightarrow (x=y \; \& \; r1=r2)]$

This is equivalent to the claim that pair-parthood has the formal property that the original Uniqueness principle attributes to $parthood_m$ itself. (I will leave it to the reader to convince him- or herself of this.)

$Uniqueness_{4P}$ tells us that no s has more than one *fusion pair* (ordered pair ⟨x, r⟩ such that x fuses s at r). Intuitively, it tells us that some things, at some locations of those things, compose no more than one further thing, at no more than one location of that further thing.

To illustrate, suppose that a statue-shaped lump of clay, Lump, fuses set s at region R, where s is a set of ordered pairs, the first element of each of these pairs being some particle of clay or other, and the second element being a region exactly occupied by the first element. Further, suppose that a clay statue, Goliath, fuses s, at some region R*. Then, according to $Uniqueness_{4P}$, Goliath=Lump and R*=R. This is what we expect from any principle that claims to be an analogue of Uniqueness. However, $Uniqueness_{4P}$ has three additional features worth noting right away.

(1) Suppose that object o exactly occupies region Ro and fuses set s there, where s={⟨a, Ra⟩,⟨b, Rb⟩}. Roughly, this means that o at Ro is composed of a at Ra and b at Rb. Now suppose that o also exactly occupies a second region, Ro* (≠Ro), and fuses set s* there. Then, given $Uniqueness_{4P}$, we can conclude that s*≠{⟨a, Ra⟩,⟨b, Rb⟩}. But this does not entail that o must have different $parts_m$ at Ro*. For it might be that s*={⟨a, Ra*⟩, ⟨b, Rb*⟩}, where Ra*≠Ra and/or Rb≠Rb*. In that case, it would be natural to say that *at Ro*, o is composed of a and b, at certain locations of those objects, whereas *at Ro**, o is 'again' composed of a and b, but at certain *other* locations of those objects. This is permitted by $Uniqueness_{4P}$.

(2) Likewise, $Uniqueness_{4P}$ permits a situation in which the very same two multi-located objects, a and b, compose different things at different locations. For we might have two different objects, o and o*, such that o fuses {⟨a, Ra⟩,⟨b, Rb⟩} at Ro, whereas o* fuses {⟨a, Ra*⟩, ⟨b, Rb*⟩} at Ro*, provided that Ra≠Ra* or Rb≠Rb*. Thus $Uniqueness_{4P}$ is analogous to a principle governing 'time-indexed' parthood that forbids a single set of things from having more than one fusion at a single instant of time, but that does not forbid a single set of things from having one fusion at one time and a different fusion at a different time.

(3) Finally, $Uniqueness_{4P}$ permits Effingham and Robson's exotic case of Brick and Wall, in which one thing (Brick), at one hundred different locations, composes a second thing (Wall), at the sum or union of those locations. Once again let Rw be a wall-shaped region exactly occupied by Wall, and let $R_1 \ldots R_{100}$ be non-intersecting brick-shaped regions whose sum or union is Rw, where each of these brick-shaped regions is exactly occupied by Brick. Lastly, let s be the set {⟨Brick, R_1⟩ . . . ⟨Brick, R_{100}⟩}. Then we can say: $Wall_{Rw}Fs$. That is, Wall at Rw is a fusion of the set of ordered pairs listed above, where the first member of each of these pairs is Brick, and the second member is one of R_1 through R_{100}. So long as this

set is fused by no more than one thing, and at no more than one region, Uniqueness$_{4P}$ is respected. It would be violated if s were fused by two different things at Rw (e.g., by a *wall* that could have had different parts and also by a 'mere mass of matter' that couldn't have had different parts), and it would be violated if s were fused by Wall at two different regions. But since there is no suggestion that either of these situations obtains, Uniqueness$_{4P}$ is not threatened by the case.

It is instructive to compare and contrast the Brick-Wall case with a version of the Lump-Goliath case, in which Lump and Goliath are stipulated to be non-identical despite both fusing the same set at the same region, R. As we have seen, the Brick-Wall case clearly obeys both WSP$_{4P}$ and Uniqueness$_{4P}$. The given version of Lump-Goliath case, on the other hand, clearly does *not* obey Uniqueness$_{4P}$. Does it obey WSP$_{4P}$? Not if either of the following is true:

LG1: $\text{Lump}\ {}_{R}{<}_{R}\ \text{Goliath}$
LG2: $\text{Goliath}\ {}_{R}{<}_{R}\ \text{Lump}$

Suppose, e.g., that LG1 is true. Then, since Lump $\neq$ Goliath, the definition Proper Part$_{4P}$ gives us:

LG3: $\text{Lump}\ {}_{R}{\ll}_{R}\ \text{Goliath}$

Together with WSP$_{4Pa}$, LG3 yields:

LG4: $\exists x \exists r^*[x\ {}_{r^*}{\ll}_{R}\ \text{Goliath}\ \&\ x\ {}_{r^*}D_{R}\ \text{Lump}]$

In words, LG4 says that some x and some r* are such x at r* is a proper part of Goliath at R and x at r* is disjoint from Lump at R. But it should be easy to see that *that* is false, given that Goliath and Lump fuse the same set at R. So, if LG1 or LG2 is true, then the Lump-Goliath case violates WSP$_{4P}$.[49] However, if those claims are false, then Lump (at R) and Goliath (at R) are not *proper* parts of each other, in which case they respect WSP$_{4P}$ (even if, roughly put, neither has a part that is disjoint from the other). So the principle that most clearly differentiates between the Brick-Wall case

49 Even in the presence of LG1 or LG2, the Lump-Goliath case still respects
QS$_{4P}$: $\forall x \forall r1[\exists y \exists r2(y_{r2}{<}_{r1}x\ \&\ (x \neq y \vee r1 \neq r2)) \rightarrow \exists y \exists r2 \exists z \exists r3(y_{r2}{<}_{r1}x\ \&\ z_{r3}{<}_{r1}x\ \&\ y_{r2}D_{r3}z)]$
For this is just the four-place counterpart of the principle that says that if a thing has a part with which it is not identical, then it has parts that are disjoint from each other (though perhaps not from the first part). Again, Lump and Goliath clearly obey this. See note 45.

and the Lump-Goliath case is Uniqueness$_{4P}$: the Brick-Wall case obeys it, whereas the Lump-Goliath case does not.[50]

So much for Uniqueness$_{4P}$. There are, of course, many other principles stated in terms of our four-place parthood predicate that have some claim to be counterparts of Uniqueness. I do not mean to suggest that these other principles are uninteresting or too distant from Uniqueness to be worth discussing. But, in light of its equivalence to the uniqueness principle governing pair-parthood, I do think that Uniqueness$_{4P}$ is the *closest* 4P-appropriate counterpart of the original principle.

5.3 Universalism

Now we are ready to formulate a counterpart of Universalism. No one should want to assert that *every* set has a fusion pair. In order for a set to have a fusion pair at all, the set must be *non-empty*, and each of its members must be an *ordered pair*, and each of these ordered pairs must be such that its first element, at its second element, is a part$_m$ of something, somewhere. But in that case, given LLP*, each such ordered pair must be an *occupation pair*, an ordered pair whose first element exactly occupies its second element. Our counterpart of Universalism, then, will say that every non-empty set of occupation pairs (every 'o-set') has at least one fusion pair. Or, in symbols:

Universalism$_{4P}$: $\forall s[[\exists y(y \in s) \ \& \ \forall y(y \in s \rightarrow \exists x \exists r1(y=\langle x, r1\rangle \ \& \ x \otimes r1))] \rightarrow \exists x \exists r1(x_{r1}Fs)]$

As our guidelines dictate, this is equivalent to the claim that pair-parthood has the formal property (over the domain of occupation pairs) that Universalism attributes to parthood$_m$.

To get a feel for Universalism$_{4P}$, let Re be a spacetime region exactly occupied by the Eiffel Tower, and let Rn be a spacetime region exactly occupied by my nose. Then Universalism$_{4P}$ tells us that if there is such a thing as the set {⟨the Eiffel Tower, Re⟩,⟨my nose, Rn⟩}, then there is some x that fuses that set at some r1. Likewise, if Re and Re* are two different spacetime regions each exactly occupied by the Eiffel Tower, then the principle tells us that if there is such a thing as the set {⟨the Eiffel Tower, Re⟩,⟨the Eiffel Tower, Re*⟩}, then some y fuses that set at some r2. The principle is silent as to *what* fuses this last set – perhaps the Eiffel

50 Thanks to Ben Caplan for discussion of this point.

Tower itself, perhaps something else.[51] This completes my discussion of Universalism and its 4P-appropriate counterpart.

5.4 Loose Ends

I will close this section by returning to issues about the interaction between parthood$_m$ and exact occupation. There are two central principles connecting these relations that virtually everyone would be willing to accept. The first is gestured at by the slogan: parts lie within their wholes. Or, alternatively, by the slogan: a whole goes at least as far as any of its parts. This is captured formally by Inheritance. It says that if x at r1 is a part$_m$ of y at r2, and if moreover r1 and r2 are both spacetime regions, then r1 is a subregion of r2.

Whereas the first principle tells us that a whole goes at least as far as its parts, the second principle tells us that a *composite* whole (one with proper parts) goes no farther than its proper parts. In other words, if a thing is composite (at a region), then any region that intersects that region will intersect a location of at least one of the thing's proper parts. Stick a pin into a location of a composite, and you will have stuck that pin into a location of at least one proper part of that composite. To express this formally, we can begin by defining an intersection predicate in terms of the subregion predicate:

Intersection: $r1INTr2 =df. \exists r3(r3 \sqsubseteq r1 \ \&\ r3 \sqsubseteq r2)$

In words, regions intersect when they have a common subregion. Next we define a compositeness predicate:

Composite: $xCr1 =df. \exists y \exists r2(y_{r2} \ll_{r1} x)$

51 Obviously, Universalism4P places no restrictions on which o-sets have fusion pairs. It allows for fusions of things no two of which ever exist at the same time. It even allows for fusions of things no two of which are located in the same spacetime (if there are multiple spacetimes). This is all very much in the spirit of the original Univeralism principle framed in terms of the two-place parthood predicate. But many philosophers will accept only a restricted fusion principle. It may be worthwhile, therefore, to see how such a principle can be stated. As a representative example, we can focus on the idea that fusion is universal for sets of things that all exist at the same instant of time. Here it will be convenient to use a new undefined predicate, 'Tr', for 'r is a global time-slice'. This lets us state the restricted fusion principle as:

Universalism$_{4Pr}$: $\forall s[[\exists y(y \in s) \ \&\ \forall y \exists r2(Tr2 \ \&\ (y \in s \rightarrow \exists x \exists r1(y=\langle x, r1\rangle \ \&\ x \otimes r1 \ \&\ r1 \sqsubseteq r2)))] \rightarrow \exists x \exists r1(x_{r1}Fs)]$

In words, this says that for any o-set of ordered pairs whose second members are all subregions of some common global time-slice, there is something that fuses this set somewhere.

That is to say, a thing x is a composite at a region r1 iff some y at some r2 is a proper part of x at r1. This lets us state the second principle connecting parthood$_m$ and exact occupation as:

Delegation: $\forall x \forall r1 \forall r2[(xCr1 \ \& \ Rr1 \ \& \ Rr2 \ \& \ r2INTr1) \rightarrow \exists y \exists r3(y_{r3} \ll_{r1} x \ \& \ r2INTr3)]$

This says that if a thing x is composite at a region r1, then any region r2 that intersects r1 also intersects a region r3 at which something is a proper part of x at r1. Together with LLP*, it tells us that if a thing x is composite at a region r1, then x *exactly occupies* r1, and any region r2 that intersects r1 also intersects some region r3 that is *exactly occupied* by something y that, at r3, is a proper part of x at r1.

One last loose end remains. I have helped myself to the predicates 'is a spacetime region' and 'is a subregion of.' The former predicate will remain undefined here. Must the latter go undefined too? I would not be deeply averse to this if there were no adequate alternative, but there seem to be two potentially viable strategies for defining 'subregion' – either in terms of 'part$_m$' and 'region', or in terms of 'subset' and 'region'. Here is the first:

Subregion1: $r1 \sqsubseteq r2$ =df. r1 is a spacetime region, r2 is a spacetime region, and r1 is a part *simpliciter* of r2,

where 'part *simpliciter*' is defined in terms of four-place parthood in accordance with the definition given in section 4. According to that definition, r1 is a part *simpliciter* of r2 just in case

$$\exists r1^*(r1 \otimes r1^*) \ \& \ \forall r1^*[r1 \otimes r1^* \rightarrow \exists r2^*(r1_{r1^*} <_{r2^*} r2)].$$

In other words, in order for r1 to count as a part *simpliciter* of r2, r1 must itself have a location (exactly occupy something), and it must be such that for each of its locations, r1*, there is some location r2* of r2 such that r1 at r1* is a part$_m$ of r2 at r2*. This may seem odd. We typically think of regions as being locations of non-regions. But do regions themselves have locations? Perhaps. I see no obvious problem with the suggestion that each region exactly occupies itself and that no region exactly occupies any other region.[52] With this suggestion in place, it is plausible that regions are

52 Is there something absurd about the idea of a self-occupying entity? While I would prefer not to say that *everything* is self-occupying (I doubt that I exactly occupy myself), I see nothing obviously implausible about the view that *some* things, such as spacetime regions, do self-occupy.

often parts *simpliciter* of other regions. So Subregion1 strikes me as tenable, at least *prima facie*. Here is the second definition:

Subregion2: $r1 \sqsubseteq r2$ =df. Rr1 & Rr2 & $r1 \subseteq r2$

Those who subscribe to this definition would presumably go on to define 'region' as 'non-empty set of spacetime points', where 'spacetime point' might then be taken as primitive. Subregion2 has the obvious drawback of identifying what appear to be concrete entities, *spacetime regions*, with what appear to be abstract entities, certain sorts of *sets*. But for those who think that they can see their way clear of this problem, the definition is available.

6. Conclusion

In sections 2-4 I argued that if multi-location is true, then parthood$_m$ is a four-place relation – one that is plausibly taken to hold between an entity x, a location r1 of x, an entity y, and a location r2 of y. When both of the 'entities' in question are material objects, it is natural to assume that their locations will always be spacetime regions, but this is not strictly required by anything that I have said. Moreover, I have left open the possibility that some of the 'entities' in question – some of the parts and/or some of the wholes – are not material objects, but rather belong to other ontological categories.[53] Still, the motivations for 4P itself and for the 4P-appropriate mereological principles discussed in section 5 are all based purely on considerations about material objects and their locations in spacetime.

Finally, I should emphasize that I have argued only for a conditional conclusion: that if multi-location is true, then so is 4P. I have taken no stance on whether we should use this conclusion as part of a *modus ponens* argument for 4P or instead as part of a *modus tollens* argument against multi-location.

53 This raises an interesting question: given that such things as states of affairs, propositions, and certain sorts of universals are notorious for their tendency to violate commonly-accepted mereological principles (Lewis 1986b and 1986c), are any of these things better behaved by the lights of our 4P-appropriate counterparts of these principles? If so, then given that 4P and the associated mereological principles can be motivated purely by appeal to considerations stemming from material objects, this might go some distance toward exonerating some of the states of affairs, propositions, etc., in question. I hope to address this issue in future work.

References

Armstrong, D. M. (1989) *Universals: An Opinionated Introduction*, Boulder: Westview.

Bricker, P. (1993) 'The Fabric of Space: Intrinsic v. Extrinsic Distance Relations', *Midwest Studies in Philosophy* 18, 271-294.

Campbell, K. (1981) 'The Metaphysic of Abstract Particulars', *Midwest Studies in Philosophy* 6, 477-488.

Donnelly, M. (forthcoming) 'Parthood and Multi-location', *Oxford Studies in Metaphysics.*

Earman, J. (1995) *Bangs, Crunches, Whimpers, and Shrieks*, Oxford: Oxford University Press.

Effingham, N. and Jon Robson, (2007) 'A Mereological Challenge to Endurantism,' *Australasian Journal of Philosophy* 85, 633-640.

Ehring, D. (1997) *Causation and Persistence,* Oxford: Oxford University Press.

Gilmore, C. (2007) 'Time Travel, Coinciding Objects, and Persistence,' *Oxford Studies in Metaphysics*, vol. 3, 177-198.

Haslanger, S. (2003) 'Persistence Through Time' in: Loux and Zimmerman (eds.), *The Oxford Handbook of Metaphysics,* Oxford: Oxford University Press, 315-354.

Hovda, P. (2009) 'What is Classical Mereology?', *Journal of Philosophical Logic* 38, 55-82.

Hudson, H. (2001) *A Materialist Metaphysics of the Human Person*, Ithaca: Cornell.

Hudson, H. (2008) 'Reply to Parsons, Reply to Heller, Reply to Rea', *Philosophy and Phenomenological Research* 76, 452-470.

Lewis, D. (1983) 'New Work for a Theory of Universals', *Australasian Journal of Philosophy* 61, 343-377.

Lewis, D. (1986a) *On the Plurality of Worlds*, Oxford: Blackwell.

Lewis, D. (1986b) 'Against Structural Universals', *Australasian Journal of Philosophy* 64, 25-46.

Lewis, D. (1986c) 'Comment on Armstrong and Forrest', *Australasian Journal of Philosophy* 64, 92-93.

Lewis. D. (1991) *Parts of Classes*, Oxford: Blackwell.

McDaniel, K. (2004) 'Modal Realism with Overlap', *Australasian Journal of Philosophy* 82, 137-152.

McDaniel, K. (2008) 'Against Composition As Identity', *Analysis* 68.2, 128–133.

McDaniel, K. (forthcoming) 'Structure-Making', *Australasian Journal of Philosophy.*

Mellor, D. H. (1980) 'On Things and Causes in Spacetime', *The British Journal for the Philosophy of Science* 31, 282-288.
Newman, A. (2002) *The Correspondence Theory of Truth*, Cambridge: Cambridge University Press.
Parsons, J. (2007) 'Theories of Location', *Oxford Studies in Metaphysics*, vol. 3, 201-232.
Parsons, J. (2008) 'Hudson on Location', *Philosophy and Phenomenological Research* 76, 427-435.
Rea, M. (1998) 'Temporal Parts Unmotivated', *Philosophical Review* 107, 225-260.
Sattig, T. (2006) *The Language and Reality of Time*, Oxford: Oxford University Press.
Sider, T. (2001) *Four Dimensionalism: An Ontology of Persistence and Time*, Oxford: Oxford University Press.
Sider, T. (2007) 'Parthood', *The Philosophical Review* 116, 51-91.
Simons, P. (1987) *Parts: A Study in Ontology*, Oxford: Clarendon Press.
Thomson, J. J. (1998) 'The Statue and the Clay', *Nous* 32, 149-173.
Van Inwagen, P. (1990) *Material Beings*, Ithaca: Cornell.

AN ALTERNATIVE TO ENDURANTISM AND PERDURANTISM: DOING WITHOUT OCCUPANTS

FRANK HOFMANN

1. Introduction

The topic of this paper is the question of persistence through time, for concrete, material particulars.

The two main contenders in the field of persistence are endurantism and perdurantism. According to endurantism, there are enduring objects: one and the same object – the numerically identical object – can be wholly present at different times. According to perdurantism, there are no such enduring objects but 'only' momentary stages, the temporal parts (which do not endure, of course), or 'only' the contents of spatiotemporal regions. So there are two versions of perdurantism, relying on temporal parts or on contents of spatiotemporal regions, respectively. The first version of perdurantism – temporal-parts perdurantism – is perhaps more common. But there is also this second version – 'content perdurantism', as I will call it – which needs to be taken into consideration. Both views, however, are significantly similar to be counted as versions of the same theory – perdurantism.

A third view has been proposed by Theodor Sider (2001) and Katherine Hawley (2001): the stage theory. The stage theory, however, is very similar to perdurantism. It lives entirely on the spirit of perdurance. And it differs from perdurantism mostly semantically, not metaphysically.[1] According to the stage view, by speaking of 'the apple' or 'the tennis ball' we refer to particular stages, whereas on 'classical' perdurantism, these expressions refer to the long-lived sums of stages (the entire persisting objects). Here, I am mostly interested in the metaphysics of persisting ob-

1 I say ‚mostly', since, as Hawley notices, semantical views may require further metaphysical assumptions. Cf. Hawley (2001), p. 43-44.

jects. Therefore, I will count the stage view as a version of perdurantism, and not as a third view.

The view that I will offer is different from perdurantism. I will call it the 'no-occupants view' (for lack of any better name). The no-occupants view does without enduring objects and without stages. This view seems to me to provide a real, genuine alternative both to endurantism and perdurantism. It differs very much *metaphysically* from endurantism and perdurantism. In the end, of course, everything depends on the exact definition of 'perdurantism' whether some theory counts as a version of perdurantism or is a real alternative to it. But the spirit of the notion of perdurantism is not to be stretched too much. Keeping this in mind, and focussing on the metaphysical issue, it seems to me that the view that I will be proposing is sufficiently different to be counted as a real, third alternative.

According to perdurantism (including the stage view), persisting objects are composed of parts which do not persist. (They may also be composed of parts which persist, but ultimately we arrive at parts that do not persist.) According to endurantism, the persisting object is enduring. According to the third view on offer, the no-occupants view, there are neither stages nor enduring objects. – What then is there? – Only the space-time regions or points and their instantiating of properties and relations. There are no objects which 'occupy' space-time. That there is a particular object at a certain spatiotemporal location is no more and no less than these spatiotemporal regions or points instantiating suitable properties (including relations). Thus, there is no need for any further entity which 'occupies' space-time – neither an enduring object nor a stage. We can do without occupants.

Here is the plan for the following. In the second section, I will briefly discuss endurantism. The third section will be concerned with perdurantism, both in the temporal-parts version and in the content version. The forth and final section will be devoted to presenting and discussing the third, alternative view, namely, the no-occupants view.

A note on methodology. I will discuss the topic of persistence within a scientifically inclined spirit. There will be no excuse for making use of the best empirical, scientific theories that we have today. And I will assume that the best scientific picture trumps everyday life assumptions, in cases of conflict. It does not make very much sense to me to try to avoid reliance on science or to bracket off all empirical assumptions. I do not offer any arguments for this methodology here (even though I think there are good arguments for it). In any case, this scientifically informed methodology is the one that I will be relying on in the following.

2. Endurantism

Let us quickly remind ourselves of the datum which is to be explained: the persistence of concrete, material objects. I take it that our datum can be stated thus:

(D) Concrete, material objects are persisting objects ('POs', for short): a PO exists at different times, it persists through time, yet it is still the same PO.

(Here, I will not be concerned with the ontology of the mental but only with non-mental material POs.)

Endurantism offers the following explanation:

(E1) POs endure, i.e., they exist wholly at different times.
(E2) POs do not have temporal parts (stages).

To say that a PO exists wholly at different times is to be taken strictly: it is the numerically identical entity which exists wholly at different times. The sameness is not of any loose sort, but is strict, numerical identity. (And this is, of course, what will lead into serious difficulties, as we will see in a moment.)

One of the most obstinate problems for endurantism is posed by the phenomenon of intrinsic qualitative change. David Lewis (1986) has offered a concise formulation of the problem. It can be restated thus: If sameness over time is strict, numerical identity, a change in intrinsic qualitative properties would require a contradiction. So if we accept the possibility of such a change, endurantism is unacceptable.

Now, many ways of dealing with this problem have been proposed as solutions. Here I will not go deeply into analysing these solutions. One solution has been to replace intrinsic qualitative properties by relations (to times). Another proposal has been to conceive of instantiation as time-indexed. Of these and other proposals in defence of endurantism, I think it is fair to say that they all suffer from one major drawback: As one can quickly realize, some quite drastic measure has to be taken in order to make intrinsic qualitative change possible without ending up in contradiction. This is enough to motivate an alternative view, perdurantism.

3. Perdurantism

Perdurantism rejects enduring objects. There are two versions of perdurantism, temporal-parts perdurantism and content perdurantism. Let us first start with temporal-parts perdurantism (which is, perhaps, more common). It can be defined by the following three theses:

(P1) POs exist partly at different times.
(P2) POs have temporal parts ('stages'); they are sums (wholes) which are mereologically composed of stages.
(P3) Stages exist wholly at a time.
(P4) POs are not enduring entities.

I will define the second version of perdurantism in due course, in section 3. For the moment it suffices to start with temporal-parts perdurantism.

Perdurantism offers an attractive picture. But it is not without serious problems. In the end, it seems to me that the problems are very significant, and it is not clear whether they can be overcome. This is why I will try to develop a third view, the no-occupants view. But first of all, let us take a look at the problems for temporal-parts perdurantism.

Two major problems arise for temporal-parts perdurantism. First, the problem of the analysis of persistence. Second, the problem of the relation between the stage and the spacetime region at which it exists.

Consider, first, the problem of the analysis of persistence. We want an informative analysis of persistence through time. According to temporal-parts perdurantism, persistence through time is to be explained in terms of temporal parts, stages. But how about the stages themselves? A dilemma seems to arise. Either the stages persist or they do not persist. (A mixed view seems to be unattractive from the very beginning.) If the stages persist, we do not have any informative analysis of persistence in terms of stages. If they do not persist, we might have an informative analysis, but then the question arises of what the relation between a stage and the time at which it exists is.

In order to answer this question, we need to take a closer look at time. We need to consider the structure of time and, in particular, the temporal extension of the 'smallest' stages. What does 'smallest' mean here? There seem to be only three possibilities. (i) Zero extension: the temporal extension of the smallest stages is zero; they are instantaneous. (ii) Quantization: the temporal extension of the smallest stages is non-zero and finite; there is a smallest quantity of temporal extension – a time quantum. (iii) No smallest units: there is no smallest temporal extension; all stages have non-zero and finite temporal extension, but they are arbitrarily small and there is no minimum and no infimum.

Now, how about the analysis of persistence? We have assumed that the stages do not persist. What is the relation between stages and times? If we opt for thesis (ii) or (iii), stages always have some finite, non-zero temporal extension. But that seems to *make them persist.* It seems unavoidable to admit that stages persist through time if it is accepted that they have finite, non-zero temporal extension. If so, we have not analysed persistence through time without relying on persistence through time, and thus we have not succeeded in providing an informative analysis of persistence. The only way out seems to be to opt for thesis (i). If the smallest stages have zero temporal extension, it seems we are no longer relying on any persistence and so we might have an analysis of persistence in terms of existence at an instantaneous time. Thus, I submit, option (i) is the most promising thesis for the temporal-parts perdurantist.

Now we can easily see what price the temporal-parts perdurantist has to pay. He is *forced* into assuming a certain structure of time, the zero-extension structure. It may very well be that this is the correct view about the structure of time – and I am not going to argue against it here. But it is quite remarkable that *as a (temporal-parts) perdurantist* one is forced to adopt this view about the structure of time. Otherwise, one is relying on persistence – the persistence of stages with non-zero, finite temporal extension – and therefore the whole point of perdurantism seems to get lost.

So far, so good. But there is another, second problem: the problem of the relation between the stage and the spacetime region at which it exists. The problem is simply this: What is the relation between a stage and the spacetime region 'at which is exists'? Either it is identity or it is not identity. If it is identity, we have eliminated the stages, since the stages just are the spacetime regions that they were said to 'exist at'. This is no longer a version of perdurantism. A view can only count as perdurantism if it endorses the claim that there are stages *over and above the spacetime regions.* It is, of course, an interesting hypothesis to deny that there are any such stages, and it will be part of the no-occupants view that this is so. But we have to realize that this is no longer temporal-parts perdurantism. The stages are supposed to 'exist at times'. And they cannot just be the spacetime regions that 'exist at their own times'.

It seems that we better had conclude that the relation is not identity. Then, there are stages, in addition to spacetime regions, and the stages are 'located at the spactime regions' (or 'exist at the spacetime regions'). But what then is this relation of 'being located at'? We might either take being-located-at as a primitive, or try to analyze it in terms of something, x. It seems quite difficult to come up with any proposal about what this x might be. So probably, one has to choose the option of taking being-located-at as an ontological primitive. Here I will stop and not go any further into discussing these matters. For, in any case, we will end up with a *dualistic conception*: according to temporal-parts perdurantism, as it turns out, there are two kinds of concrete entities, the spacetime regions and the stages which are located at the spacetime regions. In addition, probably one has to accept being-located-at as an ontological primitive. This is unattractive, it seems to me. That the concrete universe is composed of two quite different kinds of concrete entities, the stages and the spacetime regions, related by a relation of being-located-at, smells of a much too commonsensical picture. The concrete universe becomes overcrowded, one might think. Two concrete entities have the very same spatiotemporal location, the stage and the spacetime region at which it is located. (These are two concrete *particulars*, one should note. We have not touched upon their properties yet.)

In sum, the problem of the relation between the stage and the spacetime region at which it is located shows that temporal-parts perdurantism is forced into a dualistic conception of the concrete universe. The problem of the analysis of persistence is perhaps solvable, but only if time turns out to have the 'right' structure. Taken together, this is enough of 'bad news' for perdurantism in order to motivate the search for an alternative, it seems to me.

An alternative is provided by a second version of perdurantism, namely, *content perdurantism*. According to this version of perdurantism, POs are the *contents* of spacetime regions:

(CP1) POs are the contents of spacetime regions.[2]
(CP2) POs are not enduring entities.

So instead of stages, 'contents' of spacetime regions are taken to be the persisting entities. For every spacetime region, its content is reified as a concrete, material object. So again, there are two things, the spacetime region and its content.

This view is similar to what Quine says about 'events'. Events are just the concrete, material contents of spatiotemporal regions, according to Quine. And if Quine uses events in an attempt at explaining persisting objects, his view amounts to content perdurantism.[3]

This quickly leads into a problem, the 'content problem': What is it for something to be the content of a spacetime region? The PO is supposed

2 Cf. Heller (1990).
3 See, for example, Bennetts interpretation of Quine in Bennett (1988), ch. 7.

to be the content of the spacetime region. But what is the relation between the content and the spacetime region? Again, there are two possibilities. If the relation is identity, then we end up with just the spacetime regions, and the contents have in effect been eliminated. If the relation is not identity, then what other relation is it? And how are contents individuated? In particular, how many contents can there be to one spacetime region? If there is a magnetic field within a certain spacetime region, is this already a content? And what if there is, in addition to the magnetic field, also a number of electrons within the same spacetime region? How many contents do we have then? It seems fair to say that it will not be easy to answer these questions. To cut things short at this point, we can jump to the most likely conclusion: Whatever the answers may be, if there are any, we will again end up with a *dualistic conception of the concrete world*. Again, we have two different kinds of material objects, the spacetime regions and their contents. And this is not too attractive a view, I submit. Let us look for an alternative which goes beyond perdurantism altogether.

4. The No-Occupants View: A third Alternative

Let us go back to the idea of smallest units of time and see if we can get rid of the contents as occupants. Suppose we accept the following hypothesis about the structure of spacetime:

(1) Spacetime consists of (extended) spacetime regions or (extensionless) spacetime points, and these are concrete particulars.[4]

(For the following, it will not matter whether we take extensionless spacetime points or extended spacetime regions.) Now we eliminate the contents and the stages:

(2) No-occupants thesis:
There are no further concrete particulars which occupy the spacetime regions or points – no stages and no contents.

In this way we can avoid a dualistic conception of the concrete. We firmly deny that there are any particulars which occupy or 'fill' our spacetime regions or points. Thus, spacetime provides the only concrete particulars. It is the concrete, indeed, par excellence.

4 Some considerations in favor of this view can be found in Hofmann (2004).

Now, however, what are we going to do about the concrete things, the matter, as it were? We do not want to fill spacetime with stages or contents. But somehow we have to provide something which makes it the case that there are tables and trees. – The guiding idea for the no-occupants view is simply this: we can provide for tables and trees by making the spacetime regions or points *instantiate properties and relations*. Spacetime is not naked, but is instantiating suitable properties (including relations). It is 'filled' in a different way, not by further concrete particulars, but by the instantiation of properties.

Such a position is hinted at, if not expressed in full, by David Lewis in a very interesting passage:

> "[A simpler monistic conception] does away with the occupants as separate things: we have the parts of spacetime, and their distance relations are the only spatiotemporal relations. The properties that we usually ascribe to occupants of spacetime – for instance, properties of mass, charge, field strength – belong in fact to parts of spacetime themselves. When a part of spacetime has a suitable distribution of local properties, then it is a particle, or a piece of a field, or a donkey, or what have you." (Lewis 1986, p. 76, fn. 55)[5]

So the *Ersatz* occupants are properties. This solves for the problem of the relation between that which 'fills' spacetime and spacetime itself: it is simply instantiation. Because we have to accept something like instantiation anyway, and independently of the issue of persistence, this does not bestow any additional ontological costs on the proposed view. We need no further primitive, such as the relation of 'being located at' (for stages) or 'being the content of' (for contents). Instantiation of properties by spacetime elements is enough. Thus, we arrive at the following thesis:

(3) Replace the 'occupants' by the instantiations of properties and relations by the spacetime regions or points. There are certain facts, namely, facts of the kind

 spacetime region x instantiates physical property P

 spacetime region x and y stand in physical relation R to one another

 These facts account for there being POs 'at' certain spacetime regions.

---

5 Curiously, David Lewis does not really elaborate this view any further, even though he seems to approve of it and does not mention any problems he sees with it. – The last sentence of the quote is unfortunate. If the 'it' refers to the part of spacetime, then the donkey and the part of spacetime are identified. It would be better to say instead that '... then *there is* a particle, or a piece of a field, or a donkey'.

Now we can rely on a further idea which has been developed into an important doctrine since Russell, the doctrine of truthmaking.[6] Truthmaker theory provides a comprehensive account of the truth of propositions (assuming, for the moment, that propositions are the ultimate truthbearers). It rests on the intuition that the entities, or real existents, in the world make certain propositions true or false. 'Truth supervenes on being', as John Bigelow (1988) has put it succinctly. Furthermore, some of these entities are facts, i.e., the instantiations of properties by concrete particulars.[7] For the present purposes, we can assume that the properties are universals. Then, the facts are the 'combinations', or 'unities', of universals and particulars. If we build on such a view of truthmakers and facts, we can now provide truthmakers for statements about ordinary persisting objects:

(4) Truthmakers and facts:
A truth such as

(i) This glass is filled with water.

has a collection of facts as a truthmaker. These facts are the instantiations of suitable physical properties and relations by spacetime elements.

Of course, these collections of facts are in general huge, indeed, enormously rich. Many many spacetime regions or points are involved, and they instantiate hugely many properties and relations. It will be extremely difficult to describe these entities in some detail. But in principle it is clear what kind of entities they are. (And in some cases it may be rather simple to describe the pattern of properties and relations which is instantiated in a spacetime region, i.e., in 'simple' cases such as, for example, fields distributed uniformly within a certain spacetime region.)

Thus, three categories are needed: spacetime elements (regions or points) as concrete particulars, universals, and facts. Instead of having spacetime 'occupied' by further concrete particulars, we take there to be facts involving spacetime elements and suitable physical properties. (These properties can, of course, be field properties and quantities.) Thus we can provide for the 'matter within spacetime', the tables and trees. There are

6 Important landmarks for truthmaker theory are Bigelow (1988), Mulligan et al. (1994), Armstrong (1997). A more recent discussion of some views about the truthmaking relation, including a new proposal, can be found in Hofmann (2008), ch. 2.

7 This view can, of course, be found in Armstrong (1997) who calls the facts ‚states of affairs'.

the facts of spacetime elements instantiating certain properties, and these facts make propositions about ordinary objects true.

Now, in order to clarify and elaborate this view further, I will mention some potential problems that it has to face, at least *prima facie*, and try to find ways of solving them. I will bring up three potential problems. These problems are: modality, vagueness, and the problem of avoiding elimination.

The first problem concerns *modality*. There may be some mismatch between the modal features of the collection of facts which is supposed to make true the proposition expressed by a sentence like

(ii) There is a glass at place x.

On the one hand, and the modal features that we ordinarily and intuitively associate with everyday life objects like glasses, trees, and tables. Intuitively, the glass of water could be somewhere else. But the collection of facts which serves as the truthmaker for statement (ii) could not be somewhere else. Its location (if one can speak of a location at all) is fixed by the spacetime elements that are the instantiating particulars for these facts.

Two points in response. Firstly, of course, there could be some other, suitable collection of facts such that it would be a truthmaker for the proposition expressed by

(iii) There is a glass at place y.

(y different from x.) Secondly, there could be a process of moving the glass from x to y. This process can be described as a causal process of facts causing other facts such that we start from the first collection of facts which is the truthmaker for (ii) and end up with a collection of facts which is the truthmaker for (iii). So, in other words, facts are the causal relata, and they allow for causal processes, according to the laws of nature.[8] If so, it is no problem in principle to understand how there could be such a thing as the movement of one ordinary object, like the glass, from one place to another. It will of course not be the very same, numerically identical entity which is first at place x and then at place y. Endurantism is firmly rejected. But there will be enough of sameness to speak of 'the same glass'.

8 Here, I am following ontologists like Armstrong and Mellor (who calls the relata of causation 'facta', in contrast to 'facts' which he takes to be merely true propositions). Cf. Armstrong (1997), Mellor (1995).

Having said that, I leave the topic of modality. Of course, this is not a comprehensive account of the relevant modal aspects, but that would require another and more extensive investigation. Perhaps, we will have to sacrifice some modal intuitions in the end. But probably we will have to do this anyway, since science does not preserve all of our ordinary life modal beliefs.

The second problem is *vagueness*. I only mention this topic, but cannot say anything substantial about it here. Again, this would require a treatment of its own. Suffice it to say that a view which takes vagueness to reside 'only' in our representations of the world, and not in the world itself, would be quite friendly to, and compatible with, the no-occupancy view. No special problems would arise for this view, as far as I can see.

The third problem is '*the problem of avoiding elimination*', as we can call it. Let me say at the very beginning that I do not feel very confident about whether there is a real problem here and, if so, how to describe it. Nevertheless, I will try to spell out the problem as best as possible, and provide an answer.

It centers on the question whether the proposal on offer can save ordinary, persisting objects and does not lead into their elimination. (I would like to avoid elimination. So I take this to be a challenge.) One might get the impression that the view on offer ultimately, if followed and spelled out fully, leads into eliminativism about ordinary, persisting objects. The persisting object, it has been proposed, is a collection of facts, comprising facts of form: spacetime region (or point) x instantiates physical property P (as mentioned in theses (3) and (4)). Call the collection of facts which is the truthmaker for (i) collection 'c'. But can we really identify the ordinary, persisting object, e.g., the glass, with the collection of facts c? Should we not rather conclude that, on the no-occupants view, there is no glass after all? For, it is fine and legitimate to appeal to facts as truthmakers for statements like (i). But when it comes to explicitly existential propositions, such as (ii) and (iii), then what is being claimed is the existence of a certain entity, namely, the glass. Nothing but the glass itself could serve as a truthmaker for such a claim. (An entity which includes the glass could do as well, of course, as a non-minimal truthmaker. But let us leave that aside for the moment.) And this entity, the glass, cannot be identical with any collection of facts like c. Or so it seems.

Now, in response to this it seems to me to be fair to admit that one might get an impression of paradoxality when thinking of things in this way. But in the end, there is no real paradox or contradiction, and there is no need for elimination either. The impression of paradoxality is due to hidden background assumptions which can be denied consistently by the proponent of the no-occupants view. These background assumptions have to do with the semantics of true predications. Once we accept the most plausible theory of truthmakers, however, these assumptions can no longer be upheld, and have to be replaced by a more coherent and ultimately much more plausible conception of true predications. I will try to spell this out in the following.

Let us begin by taking a closer look at the objection. Consider some ordinary truths like

(5) This glass is full of water.
(6) The table is made of wood.
(7) The cat weighs two kilograms.

All of these are truth-evaluable, according to the no-occupants theorist. So suppose they are true. (We set aside any 'non-serious' or 'loose' talk, any façon-de-parler interpretation; we consider serious, literal talk which aims at a true description of reality, even if not expressed in scientific vocabulary.) This implies that there are suitable truthmakers. For example, there exists a truthmaker, some entity x, which makes the predication (5) true. But then, it seems, this truthmaker x must simply be the fact that the glass instantiates the property of being full of water. A contingent, ordinary predication is made true by the instantiation of the property (which is denoted by the predicate) by the subject of which it is predicated. And this instantiation simply is the just-mentioned fact.

At the same time, however, the no-occupants theorist wants to identify the glass with a collection of facts. But surely, this collection of facts cannot instantiate the property of being full of water. Even if there is such a property, the collection of facts does not even seem to be a possible candidate for instantiating such a property. It very much sounds like a category mistake to think that such a collection of facts could instantiate this property. At least, it is implausible to think so. The only way out, it seems, for the no-occupants theorist is to give up the identification of the glass with the collection of facts and, thus, to eliminate the ordinary object. Consequently, the ordinary predications will turn out not to be true after all.

How are we to deal with this objection? – It seems to me that there is indeed a certain air of paradox about the no-occupants view. But this impression of paradox, or implausibility, is due to certain background assumptions, and not to the view itself. Neither the assumption that there are no occupants of spacetime, nor the principles of truthmaker theory are paradoxical. It is merely due to certain additional assumptions that the view appears to be implausible, or even paradoxical. In order to make good on this reply, let us consider the matter in more detail.[9]

The answer comes in two parts. The first part clears away certain obstacles, but a problem still remains. The second part contains the solution of this remaining problem.

As a first part of the answer, let us focus on the relation between predicates and properties. What we have to recognize is that an ordinary predicate like ,is full of water' or ,is made of wood' need not denote a property. Some predicates denote a property, some don't. And if no property is denoted, the truthmaker for the corresponding predication cannot involve the instantiation of such a property. So there will be no problem with identifying the ordinary object (the glass, for example) with the collection of facts.

At this point, one may wonder whether one cannot always move easily from a predication to the existence of the corresponding property. – The answer is that true predications do not license the move to the existence of a corresponding property. From

(k) The cat is F.

it does not follow that

(l) The cat instantiates the property of being F.

Harmless as it may seem, this inference is not valid. Not because it treats instantiation as a relation (which is probably not correct). The real reason why the inference is not valid is the introduction of the existence claim about the property, implicit in the expression 'the property of being F'. Predication is one thing, but talk about properties is quite another. We may summarize this by saying: there is no apriori ascent from true predication to a corresponding property.

Now you may wonder: How about the inference from

(m) The cat is F.

to

9 If it is part of one's background assumptions that an ordinary object must be something like an enduring entity, then this will of course create pressure on the no-occupants view. But clearly, such a background assumption cannot count as a reason or argument against the view, since it would be question-begging. Perhaps, ordinary life metaphysics is more favorable to enduring entities than to perduring ones, but it is unclear how much of endurantism is really endorsed by common sense. Compare Hawley (2001), p. 11.

(n) There is something which is the cat.

Is this inference valid? Is the introduction of the existence claim about the subject equally problematic?

Ultimately, it seems to me, the answer must be, yes. The inference is not valid either. To accept such an inference is already to assume that a certain harmony, or correspondence, between language and ontology is in place. We are already assuming that the language which we are using, or which we use in order to formalize our language, is to a certain degree 'ontologically transparent'. It is ontologically transparent with respect to names (the subject position). And this is, of course, a quite common way of treating language, by now. In philosophy, we have been accustomed to treating language as ontologically transparent with respect to names. But this assumption is not obviously true for any language. And it seems to be wrong for many of our ordinary ways of using language. We speak of many things being thus and so, of possibilities and probabilities, of average people, of particular groups, pluralities, collections, of absences and presences, and so on, without taking ourselves to be committed to the existence of these 'things'. (Please note that the underlying reason has nothing to do with entering the realm or discourse of fiction. The treatment of fictional expressions, like 'Holmes' or 'Zeus', is quite a different issue. We are talking about the world, and we attempt to describe it correctly – we are not indulging in fictional stories.) We may summarize this by saying: there is no apriori ascent from true predication to the subject.

But what is the situation now? – We still have a problem, it seems. For, suppose the predicate does denote a property. For example, the predicate ‚weighs two kilograms' might be denoting a property, a certain determinate mass property. Call it 'M'. Then the paradox is that, even though the predication (7) is said to be true, the collection of facts with which the cat is identified cannot instantiate M. Spacetime points or regions can instantiate mass properties like M, but not a collection of facts comprising the instantiations of properties and relations by spacetime points or regions. So the no-occupants theorist cannot come up with a plausible position about the truthmaker of some ordinary predications. Ultimately, he is forced to give up the identification and to eliminate the ordinary object.

Now comes the second part of the answer, in reply to this problem. It will be that even in this case, the truthmaker need not take the form that we may have expected, guided by our common sense conception. Some ordinary predications are true, and have truthmakers, but even if they have a subject (such as the cat which is identical with a collection of facts) and the predicate denotes a property, it need not be the instantiation of this property by the subject which is the truthmaker of the predication. The ontological structure of the relevant truthmaker, or truthmakers, can be different; and it cannot be derived by apriori considerations merely from the surface semantics of the true statement. Even in the case we are considering, we cannot move easily to the instantiation of the property by the subject. We may summarize this by saying: there is no apriori ascent from true predication to the instantiation of the property by the subject, even granted the existence of the subject and the property; in short, there is no apriori ascent from true predication to the 'corresponding' fact.

The point bears repeating. There is a quite persistent tendency to try to read off the ontological structure of truthmakers from the surface form of sentences. This tendency must be resisted. We cannot simply derive, by merely understanding the statement in question, what the truthmaker is like, ontologically. What truthmaker, or truthmakers, a true statement has is a matter of theorizing. And such a theory must be responsive to various different considerations that, in general, extend very much beyond the statement in question.

What does this mean for the case at hand? – The no-occupants view can now be put in the following way. The true predication, like (7), has a subject (an ordinary object). This subject is identical with a certain collection of facts. Call it ‚x'. (Let us again put issues of vagueness to one side.) This is the identification claim, as part of the no-occupancy view. The no-occupants view is not eliminativist. But there is no truthmaker for the predication which is the instantiation, by this subject, of the property denoted by the predicate; there is no such fact which neatly 'corresponds' to the true predication. Nevertheless, there is a truthmaker (and there are even many truthmakers, arguably). The collection of facts, x, itself may be such a truthmaker. The facts comprising this collection, being instantiations of a property pattern by a certain spacetime region, make it the case that it is true that there is a cat and that this cat weighs two kilograms. No further entity is required. In particular, x need not instantiate property M; x itself is just enough of a truthmaker, in this case. Of course, other entities may also be truthmakers for the statement. The entire world, for example, is such a truthmaker. And various other, less comprehensive truthmakers may exist. This is just the well-known and generally accepted non-uniqueness of truthmakers. A given true statement may have many different truthmakers. But we can focus on one particular truthmaker, the collection x of facts. This entity, x, seems to be entirely sufficient as a truthmaker for the predication (7).

How can it be that x is a truthmaker for (7) without its instantiating property M? – In order for x to be a truthmaker for (7), x has to be a certain way or somehow. But this need not amount to x's instantiating property M. That x is a certain way (or is somehow) may amount to no more than x's being a collection of facts each of which is a certain way. And that a fact belonging to this collection of facts is a certain way amounts to no more than its being the fact that it is, i.e., its being the instantiation of a property by a spacetime point or region. If the right properties are instantiated by the spacetime points or regions, then there are the right facts – then there are facts which are a certain way. The collection of facts needs not instantiate a further property in order to be a certain way and, thus, to make true a certain statement.

No fact in the collection x of facts is such that it is an instantiation of property M. And there need not be any such fact in this collection. Nothing instantiates this property M, and nothing needs to instantiate it in order for the predication to be true. – Still some air of paradox?, you may wonder. – But we have to keep in mind that the fact that the particular predicate 'weighs two kilograms' denotes a property (as we have supposed) is irrelevant. If we switch to some other predicate which does not denote a property – perhaps, the predicate 'is full of water' – then the tendency to demand the instantiation of a property corresponding to the predicate loses its grip on us. Just as there is no apriori ascent to the property (and no apriori ascent from true predication to the subject), there is no apriori ascent to the instantiation of the property by the subject.

By now, we have solved the problem (if indeed there was any problem at all, of which I am not sure). Background assumptions about what entities, or what sort of entities, are required for the truth of predications have been at work in generating an impression of implausibility or even paradoxality – ultimately creating the alleged eliminativist conclusion. We have to give up these background assumptions, since they do not provide the best view of the relation between sentences and the world. A much more plausible – and systematic – view of the relation between sentences and the world is provided by truthmaker theory. The example at hand is a good case for seeing how truthmaker theory does very fruitful work and helps to decide ontological debates. In particular, we can identify the ordinary, persisting object with a suitable collection of facts. There is no need for elimination. To this extent, we can save the everyday life conception of the world. And we can do so without introducing problematic entities like stages, contents, or even enduring entities. This, I take it, makes a good case for the no-occupants view.

References

Armstrong, David (1997) *A World of States of Affairs*, Cambridge: Cambridge University Press.

Bennett, Jonathan (1988) *Events and Their Names*, Indianapolis: Hackett Publishing Company.

Bigelow, John (1988) *The Reality of Numbers*, Oxford: Clarendon.

Hawley; Katherine (2001) *How Things Persist*, Oxford: Clarendon.

Heller, Mark (1990) *The Ontology of Physical Objects*, Cambridge: Cambridge University Press.

Hofmann, Frank (2004) 'Substrate', *Metaphysica* 5, 35-62.

Hofmann, Frank (2008) *Die Metaphysik der Tatsachen*, Paderborn: mentis Verlag.

Lewis, David (1986) *The Plurality of Worlds*, Oxford: Blackwell.

Mellor, Hugh (1995) *The Facts of Causation*, London: Routledge.

Mulligan, K., Simons, P., Smith, B. (1994) 'Truth-makers', *Philosophy and Phenomenological Research* 44, 287-322.

Sider, Theodore (2001) *Four-dimensionalism: An Ontology of Persistence and Time*, Oxford: Oxford University Press.

THE BEGINNINGS OF GREEK ONTOLOGY AND THE QUESTION OF TEMPORAL BEING

JAN SZAIF

1. Introductory Remarks

Being became a topic of Greek philosophy with Parmenides, and already in Parmenides the question of temporal existence, or being as persistence through time, is a central concern—yet only negatively, so to speak, since Parmenides' tries to demonstrate that we cannot maintain the idea of objects whose existence is temporal, since the notion of temporal being presupposes the concepts of coming-to-be and passing-away, but these concepts are inconsistent.

The part of my paper which is strictly historical will focus on Parmenides. In my concluding remarks, I will briefly compare Parmenides' position to the developments in Democritus, Plato, and Aristotle—other major Greek philosophers who make use of the notion of being—in order to give some indications as to how they moved forward toward a theory of being that accounts for persistence through time. Yet before I talk about Parmenides and reactions to Parmenides, I will first comment on the most important semantic functions and connotations of the Greek word for 'to be' (*esti*) and on how these semantic functions and connotations were incorporated into the philosophical notion of 'being'. In doing so, I will take into account some of the theoretical developments regarding the meaning of 'to be' that we encounter in ancient and in scholastic philosophy—to the extent that this helps us to better understand the various semantic functions of "*esti*".

As a preliminary, I also want to mention a crucial difference of approach between ancient Greek and contemporary ontology. (The word 'ontology' is actually a coinage of the early modern era,[1] yet it was introduced as a designation for the Aristotelian concept of an investigation of

1 Cf. K. Kremer / U. Wolf, "Ontologie", in HWPh, vol. 6, 1189 ff., Basel 1984.

being qua being. So when I refer to the classical understanding of ontology, I mean the classical project of an investigation of being qua being.) Today, the dominant meaning of this word seems to be defined by Quine's question regarding "what there is".[2] According to this question, ontology as a theory is an account of what types or categories of objects are part of our theory of the world and what their formal characteristics and relations are. Ancient ontology, to be sure, also addresses these questions, but it starts with a different question whose focus is directed to the attribute of being itself. As Plato put it in the *Sophist* (243d-244b): Before we construct theories about the basic elements or features among the 'things-that-are' (i.e. reality), we ought first to clarify what this attribute itself, 'to be', is. Plato complains that his predecessors have produced theories about what kinds of basic beings there are, but have failed to clarify this very concept of 'being', so that, in a way, they didn't yet know what they were talking about. Aristotle follows Plato in this respect when he formulates the idea of a most universal and foundational form of theoretical investigation which would be an investigation of being qua being (i.e. an investigation of that-which-is insofar as it is; *Metaph.* IV.1). For he not only singles out the attribute 'to be' as *prima facie* the most basic and foundational attribute of all, which can be used to define the scope of a most general and foundational type of investigation, but he also hastens to emphasize that such an investigation cannot get started as long as we haven't investigated the various significations of the word 'to be' (*Metaph.* IV.2, 1003a33-b19; cf. VII.1, 1028a10 *ff.*; VI.4, 1027b29 *ff.*; V.7).

Thus classical Greek ontology is focused on the very concept 'to be' in a way unknown to contemporary analytical ontology. From the viewpoint of ancient ontology (and also from that of its medieval and early modern receptions) this investigation of the concept 'to be' is also more than just a preliminary step needed to prevent fallacies of equivocation. Rather, the results of this investigation are supposed to provide some essential starting-points for ontological theory or metaphysics. Now, this concept 'to be' might just not have the importance for metaphysics that the ancients attributed to it, which would justify the modern neglect of this kind of question. I am inclined to believe that a defense of the ancient approach is possible, but won't take a stance on this question in this paper.[3] Suffice it to say that the way in which Heidegger tried to revive the

---

2 Cf. W. V. O. Quine, *On what there is*, in *From a Logical Point of View*, revised ed., Cambridge (Mass.) 1980, 1-19, 1.

3 Cf. J. Szaif, *Der Sinn von „sein". Grundlinien einer Rekonstruktion des philosophischen Begriffs des Seienden*, Freiburg/München 2003; *idem*, *Plädoyer für eine formalsemantische Rekonstruktion des philosophischen Begriffs des Seienden*, in M. Lutz-Bachmann, Th. M. Schmidt (eds.), *Metaphysik*

'question of being' has obviously failed to provide a viable program for sustained philosophical research.

2. One-place and 2-place uses of *"esti"*

2.1 A traditional distinction of 'surface grammar': 'copulative' versus 'substantive' use (1-place or 2-place construction)

The "*esti*" can be used like a verb that is predicated of a subject. Whatever the underlying logical structure, at least from the point of view of surface grammar it can be treated as a 1-place predicate. In English, such a use of "is" as a 1-place predicate (as in "He is no more") is rare, while it is a well established usage in ancient Greek. Its meaning in Greek is either existential or 'veridical' (to use a term introduced by Charles Kahn into the discussion of the meanings of the Greek verb 'to be'[4]). In the latter case the "*esti*" is predicated of the content of an assertion or belief and indicates the truth of this belief or assertion or the obtaining of the corresponding state of affairs.

The most common use of "*esti*" has a 2-place construction which requires a supplementation by some predicate term (or rather helps to form a predicate term, as in "... is heavy"). Now, from the logical point of view, this 2-place construction can have a different logical import depending on what kinds of terms supplement the *esti* in the first and second position. If both terms are singular, we have a statement of identity; if they are both general and the first is supplemented by a universal quantifier, we may be dealing with a case of subordination; and if the first is a singular and the second a general term, it is a case of subsumption. (Yet it may also be a case of subordination, if we have an abstract singular term in the subject position.) These distinctions become important when we are dealing, for instance, with Platonic dialectic. Yet first of all we have to note that neither Plato nor Aristotle distinguish between different meanings of the 2-place "is" although they seem perfectly able to distinguish between, say, a statement of identity and a statement of subsumption. Also among modern linguists and philosophers of language we find the view that the 2-place "is" does not have different meanings in these different types of

heute — Problem und Perspektiven der Ontologie / Metaphysics Today — Problems and Prospects of Ontology, Freiburg/München 2007, 146-195.

4 Ch. Kahn, *The Verb 'Be' in Ancient Greek* (= W. M. Verhaar [ed.], *The Verb 'Be' and its Synonyms, part 6*), Dordrecht 1973; idem, *On the Theory of the Verb "To Be"*, in M. K. Munitz (ed.), *Logic and Ontology*, New York 1973, 1-20.

statement.[5] According to such a view, it is the logical form of the sentence as a whole, not the meaning of the "is", that distinguishes the identity statement "Tullius is Cicero" from the predicative statement "Tullius is brilliant".

The "is" or *esti* in the 2-place construction has traditionally been named the "copula" since it *links* two terms together, with or without the addition of a quantifier and/or a negation sign. In contrast to this *verbum copulativum*, the 1-place construction was named the *verbum substantivum* by older lexicographers, since it can be used to assert the existence of some substantial entity which is then, in turn, the bearer of properties that can be predicated of it with the help of the copula. (This sort of terminology has obvious Aristotelian roots, even though it greatly simplifies a much more complex understanding of the issue in Aristotle.)

2.2 One-place uses of "esti"

2.2.1 Existence

The Greek *esti* is frequently used similarly to the English "there is", often in the upfront position of a sentence—like its English counterpart— and followed by a descriptive phrase. This use corresponds quite well to the existential quantifier in modern logic. We can call this the notion of *quantificational existence* (since it essentially involves the quantifier "at least one"). The hallmark of our modern understanding of existence as a form of quantification is that existence should either not be conceived as a predicate at all or, at least, not as a first-order predicate, which means that neither existence nor non-existence can be predicated of any particular object. Using Russell's theory of definite descriptions, sentences which deny the existence of some fictitious object ("There is no Santa Claus;" "Santa Claus doesn't exist") will be analyzed as denying that the descriptive content associated with the name "Santa Claus" is satisfied by one (and only one) object in the world. Many would hold that even a sentence with a genuine singular term that cannot be resolved into a definite description could still be reduced to a general statement of existence by means of the identity sign: $(\exists x)\ (a=x)$. Accordingly, "John exists" would mean: "There is somebody who is John." (One could also say that 'to be identical with John' is just another kind of definite description.)

---

5 Cf. Kahn, *The Verb 'Be'* ... [fn. 4], 400, fn. 33; C. J. F. Williams: *What is Existence?*, Oxford 1981, 10-12; J. Lyons: *Introduction to Theoretical Linguistics*, Cambridge 1968, 322 f.

Since a general assertion of existence is equivalent to the claim that a certain predicative content is true of, or satisfied by, some object or objects, it needn't come as a surprise that the scholastics subsumed the notion of being linked to general assertions of existence under the heading of veridical being.

There is another type of use of the existential *esti* in Greek which seems closely related to the idea of living (as in the English phrase "He is no more"). Some have claimed that in this case, "*esti*" simply means "to live", based on the observation that this specific usage is normally found in connection with humans or personified entities.[6] Yet this may simply be a consequence of the fact that the by far most relevant case of individual temporal existence for us is the case of human temporal existence (or also that of certain animals with whom we interact at a quasi personal level, like pets, or certain institutions that have a high emotional value for us). Moreover, if the 'to be' simply *meant* 'to live' when predicated of living beings, then the famous statement of Aristotle's that for a living thing *to be* is *to live* (*De anima*, II.4, 415b13) would turn out to be a mere tautology. Aristotle's point is rather that in the case of living beings, continued temporal existence (i.e. persistence) is living. This is not a claim about the *sense* of "to be" in such a context but, rather, a claim about what kind of fact constitutes temporal existence in the case of living beings.

There is a good case to be made that existence in this specific sense is a first-order predicate of objects. It certainly does not involve the philosophical puzzles of non-existence that provided one of the motives for the theory of definite descriptions, since if a particular continuant ceases to exist, this does not imply that its name looses its reference.

Matters would be relatively easy if we could simply distinguish between these two types of existence, i.e. existential quantification on the one hand and assertions of individual temporal existence (persistence) on the other. Yet our language also needs tools to distinguish between real objects on the one hand and fictitious (or, more broadly speaking, mind-dependent) objects on the other, since we have the ability to invent objects or persons and are also liable to erroneously believe in, or hallucinate, the existence of objects or persons and to relate to them as if they were real. The ability to make this distinction is in fact part of our basic cognitive and linguistic abilities. It should not be confused with ordinary assertions of existence: For instance, when I am asking a group of bystanders: "Is there somebody who can help me?," I am not asking whether or not the person who can help me, or the class of persons who can help me, is a

6 Cf. Kahn, *The Verb 'Be'* ... [fn. 4], 244 f., 398 f.

fictitious or unreal person or a fictitious or unreal class of persons (like Santa Claus is, or unicorns are).

There can be no doubt that the ancient Greek language uses the term 'to be' in order to assert or deny the reality of certain objects, for instance the chimaera.[7] Yet philosophical theories about the semantic import of such assertions differ widely. One option is to reduce such assertions to general statements of existence that involve a definite description. If, on the other hand, you want to allow for genuine reference to fictitious objects, then one option for your ontology would be to introduce a conception of intentional objects such that those intentional objects can either exist in reality or not exist. In this case, we would want to distinguish the notion of reality (*esse in re*) or real existence from a weaker sense of existence that can be captured by the notion of existential quantification, since once we have allowed fictitious objects into our ontology, they can also form a universe of discourse and serve as values for the variable in an existential quantification. (In other words, each of them would also count as *something* about which some true or false assertion can be made.)

Now, let's assume that our theoretical account of what is going on in the case of apparent reference to fictitious objects rejects the idea of genuine reference to such objects (perhaps because we adhere to a causal theory of reference). I think that *even then* there would still be a philosophical need for a concept of real existence in order to express that very point that genuine reference can occur only in relation to a *real* object. Note also that the analysis of singular statements of existence through the logical form $(\exists x)\ (a=x)$ is not able to capture this specific aspect, since *if* we could refer to a fictitious object *a*, then it would of course also be true to say that *something* is (is identical with) *a*—which satisfies the form $(\exists x)\ (a=x)$. In other words, this sentence form is neutral as to the distinction between real and non-real objects. Moreover, statements like "Unicorns don't exist" do of course presuppose that the universe of discourse is restricted to objects of the real world. Thus even in the case of assertions of general existence which take a stand with respect of a fictitious *class* of objects, the notion of reality (*esse in re*) is presupposed.

To be sure, one could avoid the terms "existence" or "being" in this connection altogether and use a word like "real" instead. Yet we are here analyzing the Greek terminology of being, and it is at a later stage in the history of philosophy that words like "reality", and even "existence", were introduced as technical terms of philosophical discourse and then gradually adopted into the vernacular. In Greek, the non-technical term 'to be'

7 Cf. Gorgias 82B3 (Diels-Kranz), n. 80 (from Sextus' quotation/paraphrase of Gorgias' dialectical treatise *On not-being*).

had to fulfill this function. (Note that the ancient Greek language still shows a close connection between this use and the locative-existential use of 'to be'.)

In our common parlance, the term "to exist" is probably more typically used to convey the connotation of what I am calling real existence (viz. in sentences like "Santa Claus doesn't exist" or "Unicorns don't exist"). Some philosophers suggest that we should distinguish between the "there is" and the verb "to exist", yet the semantic boundaries of these expressions in our ordinary language are not sharply drawn,[8] and thus I find it preferable to speak of "real existence" for the sake of disambiguation.

There is another important philosophical area for which we need a notion of real existence, namely when existence is discussed with respect to an entire ontological class of objects—the sort of question that was dubbed 'external questions of existence' by Carnap[9] and others. A typical example from Plato for this sort of question is the discussion as to whether or not only bodies are real (discussed in Plato's *Sophist* under the heading of *being*—i.e. whether or not only bodies *are*). And of course it is also a central concern for Plato, the Platonic Academy, and Aristotle, whether or not intellectual Forms and, more specifically, the objects of Euclidean geometry (either as Forms or as a separate ontological category of mathematical entities) are real only qua properties of material objects or also independently of such instantiation.

While the question of the reality or unreality of an object of (what might turn out to be) a delusion is an empirical question, such 'external' questions of existence cannot be decided merely on an empirical basis but are genuine philosophical questions. They are also not part of our pre-theoretical concerns. Yet our ability to make sense of questions of external existence is, I submit, based on our familiarity with internal questions of existence which do matter in our everyday discourse.

What kind of criterion could help with external questions of existence? If one does not follow Carnap in holding that such questions can ultimately be decided only based on certain pragmatic considerations, we would need a strictly theoretical criterion. It would of course be question-begging to say that real objects are objects that exist independently of our mental acts, for we would then still need a criterion for determining what counts as mind-independent existence. (For instance, what would be the criterion for attributing reality to Platonic Forms, or to Fregean thoughts?

8 Cf. M. Reiche, *Nonexistent Objects*, in *Stanford Encyclopedia of Philosophy*, 2006, http://plato.stanford.edu/entries/nonexistent-objects/

9 Cf. R. Carnap, "Empiricism, Semantics, and Ontology," *Revue Internationale de Philosophie* 4 (1950), 20-40.

It does not seem enough simply to *stipulate* that they exist independently of being conceived by us.)

External questions of existence can be raised not only regarding objects, but also with respect to features or properties. The scholastic terminology that was based on the Aristotelian and Neoplatonist traditions distinguished between mere objects of thought and 'real' objects (*ens rationis vs. ens reale*) and included within the broad, negatively defined, class of *entia rationis* not only fictitious objects but also certain kinds of characteristics that do have a foundation in reality although they are still not 'real' in the required sense. *Negative* characteristics and many kinds of *relation* were viewed as such non-real characteristics. Leaving aside the more complicated case of relations, the standard example for a *negative* characteristic is that of blindness. Blindness is a *privation* of sight (understood as a faculty). Beings that are naturally able to acquire the faculty of sight are called blind when they fail to acquire it or when they lose it. Now, the lack of sight in a blind person or animal is, of course, nothing fictitious but a factual condition of a real object. Yet, still, what corresponds to our predicate "blind" in reality, is not a real property but rather the lack of such a property in a being that is naturally apt to having this property (while it would not be ontologically adequate to say of sight that it is a lack of blindness). Note that this sort of distinction (which can be traced back to Plato, possibly even to Parmenides, and receives its first systematic formulation in Aristotle) presupposes objective criteria for distinguishing between positive and negative attributes, which in many cases turns out to be much more problematic than in the intuitively plausible example of blindness.[10]

In order to distinguish between the sense of existence that can be captured by the existential quantifier and the sense of existence that involves a notion of *real* existence, I will also use the terminology of *veridical* versus *positional* existence (veridical existence, since it requires no more than that a certain descriptive content is true of something; *positional* existence, since something has been posited as a mind-independent entity—as part of the furniture of the world, so to speak). As we have seen, there are two very different respects in which a distinction between veridical and positional existence may recommend itself to the philosophical investigator. The first relates to the case of particular objects that are conceived after the model of real objects but are not real (fictitious objects); the second concerns entire ontological classes of objects which may turn out to lack real being.

The notion of real existence (*esse in re*, positional being) should not be identified with that of actual existence, which would be a formal attribute

10 Cf. G. Frege, "Die Verneinung. Eine logische Untersuchung," in *Beiträge zur Philosophie des deutschen Idealismus* 1 (1918/19), 143-157.

of *possibilia* that happen to become actualized, since theories are conceivable (and were already discussed in the Middle Ages--not to mention David Lewis) according to which *possibilia* are real objects even if they are not actualized (part of the actual world). Furthermore, neither real nor actual existence is conceptually identical with singular temporal existence (persistence) since it is conceivable that an ontology includes *non-temporal* particulars (which, by definition, cannot *persist* since they are not subject to time). Moreover, continuants in possible worlds also persist, even though they are not actual.

2.2.2 *"esti"* as a verb of truth ascription (veridical use)

A particular feature of the Greek "*esti*" is the veridical use. One can use "esti" to assert the truth of some propositional content or the obtaining of a state of affairs. To give an example of how this looks in Greek: A typical sentence involving a veridical *esti* would have the form: "What you are saying, *esti* ('is', i.e. 'is true' or 'is the case')— *esti tauta ha legeis*. Another noticeable usage that is based on this sort of construction is the use of the participle *onta* ("(things) that are") as an object with a verb of saying or believing. In this position, *onta* can be exchanged synonymously with the adjective "true" (*alêthê*) and used to indicate that something which has been said is true, or obtains. (Note that we are not talking here about attributive uses of "true", as in "true gold" or the like, but about propositional truth.)

Against Kahn[11], I maintain that this construction is not derived from the comparative veridical construction "This is such as you say" (*esti tauta houtôs hôs sy legeis*). The bearer of the veridical meaning in the latter construction is not the *estin*, but the *estin houtôs* ("It is so"). Both constructions are important, the latter because it supports the idea that truth is some sort of agreement or correspondence between words and the world. Yet the veridical use of the participle is based on the former construction which does not include an explicit or implicit comparative clause, and which can easily be conflated with an existential use.

Since in this construction the veridical *esti* seems to be predicated of something, one wonders what this is. Is it a state of affairs, a propositional content, a sentence utterance, or a belief? Now, if one wants to answer this question, one is already engaged in semantic and ontological theory. Especially at the beginning of the development of theories about the truth

11 Cf. Kahn, *The Verb 'Be'* ... [fn. 4], 331 ff., and my critical comments in Szaif, *Platons Begriff der Wahrheit*, Freiburg/München 1996, 42-49.

bearers we cannot expect to encounter the theoretical notion of propositional contents (understood as a specific kind of abstract objects). In Greek language and philosophy, especially in its earlier stages—as can be gleaned from Plato's use of the Greek words for "truth" and "true"[12], but also from other older sources—truth seems to have been viewed primarily as something located at the world level, as some aspect of reality that we can come to know, or get deceived about, or as something adequately, or distortedly, depicted in statements. Less frequently, truth was also attributed as a property to assertions and beliefs which are such as to present reality, or some real object, the way it actually is. We find a similar ambiguity in the use of the veridical "is".[13] Now, there are very good philosophical reasons for claiming that truth is really, at least in its primary sense, a property of propositional contents and that those contents ought to be classified as a specific type of abstract objects and distinguished from sentence types and beliefs. Yet this would be a philosophical discovery. One would of course claim for this discovery that it is founded on the rules of language and thus rooted in our implicit understanding of the language. Yet this reference to our implicit understanding has to be distinguished from the question as to how the naive language user, or the initial theorizer, would interpret their linguistic practice.

A related question is whether this veridical usage is clearly distinct from existential uses from the point of view of the naive language user or initial theorizer. All the major Greek philosophers seem to have been temporalists regarding the truth value of statements and beliefs about changeable objects. There seem to be rather vague borderlines between the temporal obtaining or temporal existence of a state of affairs, the taking place of a certain kind of event at a certain time, the spatial constellation of certain objects in relation to each other at a certain time, the temporal presence of a certain characteristic in an object, and finally the temporal existence of an object. From the point of view of the Greek language user (and the initial theorizer), this may all somehow fall under the scope of a broad notion of *onta* (that 'what is') which covers not only clear cases of temporal veridical being, but also being in the sense of real existence, and all those cases that lie somehow in between (like events and spatial constellations). The best evidence for this broad notion of being is the famous Homeric formula for the seer or prophetic man as knowing

12 Cf. Szaif, *Platons Begriff*... [fn. 11], 25-71.

13 Thus we find in Aristotle two theories of veridical being, one that seems to treat it as a property of thoughts, another that attributes it to the 'truthmakers' of affirmative thoughts (to use a modern jargon). This undecidedness reflects, I think, an ambiguity in the Greek language of truth and veridical being. Cf. E. Tugendhat, *Der Wahrheitsbegriff bei Aristoteles*, in *Philosophische Aufsätze*, Frankfurt a. M. 1992, 251-260.

"that which is, will be, or was in the past" (e.g. Il. 1.69-70). (Incidentally, I cannot detect any substantial difference between the meaning of the future particle *esomena* in the prophetic formula just quoted and a certain usage of "will be" in English. When we sing along with Doris Day in questionable Spanish but impeccable English: "Que sera, sera—Whatever will be, will be," the "will be" has the same scope as the veridical *esti*. Thus we encounter here a trace of the veridical use of 'to be' in English.)

Let's take stock: Our considerations on the semantics of 1-place "*esti*" have the result that the main usages are veridical and existential. Depending on the context, the existential use can acquire a specific connotation of *real* existence (which I will also call positional being). There is, moreover, a use that serves specifically to express the temporal persistence of a continuant. The veridical and existential usages need to be distinguished when we read and translate Greek texts, yet we have also indications that their boundaries were not sharply drawn, which meant that once philosophers had started to think about the implications of the uses of "*esti*", there was some temptation to understand assertions of veridical being simply as an assertion of real existence with respect to the content of the assertion. This tendency was reinforced by the fact that the Greek word for truth (*alêtheia*) can also mean reality.

2.3 Entailment relations between the 2-place and 1-place uses

I have talked about veridical and existential uses of the substantive "esti" as they are rooted in the ancient Greek vernacular. As far as the copula is concerned, it is particularly relevant for our purposes to get clear about its veridical and existential connotations, which derive from certain entailment relations between the 2-place and 1-place uses. I will comment on these entailment relations based on the material that we find in Plato and Aristotle, yet I think that their usage, in this specific regard, reflects the common understanding of the semantic import of the copula.

In talking about 1-place uses, I will speak of *prima facie* existential uses, except where the *esti* is predicated of a propositional content (which would be a veridical use). I will call them 'prima facie' existential, because we do find instances of 1-place usage where this use clearly is elliptical for "to be something or other", i.e. based on the copula.

Let me explain this briefly: In a famous passage in the *Republic* (476e-480a), Plato argues that there are objects of cognition which both *are and are not* and which therefore do not allow for a firm cognitive grasp. What he means by "are and are not" is at first rather puzzling, but as soon as he gives examples we realize that he is talking about predicative being: to be

F and not to be F. Depending on the context of judgment, we can call the same object both large and also, in a different context, not large, or small. Thus the mixture of being and not-being is here a mixture of contrary modes of being to be expressed by a copulative "to be" (e.g. "to be large and not to be large"). It does not suggest a (paradoxical) mixture of existence versus non-existence. In such a case, we can call the 1-place use *elliptical*, following a now common practice among scholars.

It is a moot point among scholars to what extent the 1-place "...is" in Plato and Aristotle is used as an *elliptical copula*. I am favoring an approach according to which the 1-place *"esti"* in these authors is often genuinely existential, yet such that there is also a systematic connection between the existential and copulative functions of *"esti"*. This connection can be spelled out in two directions, which I will call the predicational implication of the existential "*esti*" and the existential implication of the copula:

2.3.1 The predicational implication of the existential "esti":

It is by now widely recognized[14] that in Plato and Aristotle a sentence of the form "a is" (where 'a' is a term that refers to a concrete or an abstract entity) always entails that there is some (singular or general) term *t* such that *a is t*. One way of putting this is to say that 'to be' (*einai*) always entails 'to be something' or 'to be something or other' (*einai ti*) [a is $\rightarrow(\exists t)$(a is t)].

Its contrapositive is equally important: If we assumed (*per impossibile*) that some x is nothing at all, this would entail that x is not (*simpliciter*), i.e. that x does not exist.

Now, the expression 'to be something' is itself ambiguous between statements of identity, subordination, or subsumption. Depending on what the subject term is of which *being something* is asserted, and on what the term is that would replace the 'something', the statement can turn out to involve subsumption, subordination, or identity (which, in a case where the subject term refers to an abstract object or universal, may also be viewed as a case of definitional equivalence). In a way it does not matter what kind of statement it is, since the crucial first step is to acknowledge that there is no way in which a thing can simply *be* without also being some-

14 Cf. the influential papers by L. Brown, "Being in the *Sophist*: A Syntactical Enquiry," *Oxford Studies in Ancient Philosophy* 4 (1986), 49-70; idem, *The Verb 'to be' in Ancient Greek Philosophy*, in S. Everson (ed.), *Language (=Companions to Ancient Thought, vol. 3)*, Cambridge 1994, 212-236. See also Ch. Kahn, "A Return to the Theory of the Verb be and the Concept of Being," *Ancient Philosophy* 24 (2004), 381-405; F. Leigh, "The Copula and Semantic Continuity in Plato's Sophist", Oxford Studies in Ancient Philosophy 35 (2008).

thing or other, i.e. without standing in relations of identity/definitional equivalence, subordination, or subsumption to a predicative content.

We might also call this the *expandability thesis*:[15] An ascription of being always has to be 'expandable' into an ascription of a specific kind of being, i.e. an ascription of some predicative content other than just 'to be'.

Now, this feature of expandability could of course also be interpreted in line with the elliptical reading of 1-place being. Lesley Brown's initial formulation of the now widely accepted expandability thesis was in part a critique of the overuse of the notion of an elliptical usage of "*esti*". Yet the real dispute is over whether or not this expandable "is" has an existential import (as it clearly has in Aristotle's *An. post.* II.1). Whether or not we call it an elliptical usage, as long as we restrict the import of 1-place being to mere determinateness (i.e. to the fact that the object in question has certain characteristics), it amounts to the same thesis as the elliptical construal. Yet there are good indications that this 1-place use, notwithstanding its 'expandability', often is meant to carry an existential import, and that is what I am going to assume here.

2.3.2 The existential implication of the copula:

The reverse entailment would be that from 2-place to 1-place being. This entailment would be trivially true if the 1-place use were only the quasi elliptical one that does not imply an existential claim. If, however, it is meant in the existential sense, then it would amount to the claim that *only* things that exist can also have any attributes or descriptive content. Plato and Aristotle seem to hold that view, at least with respect to non-intentional predicates.

2.3.3 Veridical implications:

We also need to mention some entailment-relations between the veridical use and the copulative use. If veridical being is asserted of some propositional content (as in "What you are asserting, is [the case]"), then this instance of veridical being can be 'unpacked' or unfolded by articulating the propositional content.[16] Ancient philosophers are focused on simple assertoric statements that very often have the form "S is P", and in such cases veridical unfolding means that, whenever it is said of some asserted

15 I am using here an expression employed by Stephen Menn. The position as such was first developed by Lesley Brown in recent literature (see above, fn. 14).

16 Cf. Ch. Kahn, "Some Philosophical Uses of 'to be' in Plato", *Phronesis* 26 (1981), 105-134.

content that *it is*, this "is" can also be interpreted as standing proxy for a copula. (In dealing with ancient texts on being, it is often hard to decide whether or not the "is" in question is strictly veridical or rather an elliptical copula with a veridical connotation.)

If we, moreover, combine the veridical unfolding with the existential implication of the copula, we get the result that, in typical cases, the truth or veridical being of an asserted content entails the existence of the object named by the subject term of the assertion.

Looking at the copula, we also need to highlight its specific veridical implication: "S is P" is equivalent to "It is (the case) that S is P". (In Greek, one would rather change the word order from "S is P" to "Is S P." This is an impossible word-order for an assertoric sentence in English, but in Greek it serves to express the specific veridical entailment of the copula.[17])

Deflationary theories of truth use this sort of equivalence as a support for their claim that truth is not a genuine (second-order) property. But one could also draw the opposite moral from this observation, *viz.* that predicative assertions all entail the ascription of the truth property to the asserted content. In the Greek philosophy of being, especially in Plato, the veridical implication of the copula is a very important feature of the concept of being.

3. Basic features of the philosophical notion of 'being' (*on, einai, ousia*)

Next I want to talk about how these different semantic functions of "*esti*" became relevant for the philosophical notion of 'being'. We want to get a better grip on what the talk of 'being' in ancient Greek philosophy means.

3.1 'being' as count noun and as uncountable noun, and its Greek counterparts

First, I will briefly address the grammatical difficulty of rendering the Greek terminology of being into English.

Greek is much more flexible than English in forming nouns from the verb "*esti*" or "is" and in expressing their grammatical relations.

The three main forms of expressing the notion of being as a noun are (1) the infinitive used as a noun (*to einai*), (2) the participle used as a noun

17 Cf. Aristotle, *Metaph.* V.7, 1017a31-35.

in the singular or the plural (with or without article, *to on*, *on ti*, *(ta) onta*), and (3) the abstract noun *ousia*. In English, one often has to use relative clauses as a translation in order to avoid ambiguity, for instance "that which is" for "*to on*".

Regarding the English noun "being", one has to distinguish between its use as a count noun and as an uncountable noun.

The *count noun* (as in "a being", "beings") is equivalent to the use of the Greek participle *on/onta* as a noun, while the *uncountable noun*, which is derived from the gerund (the 'being' of something, or 'being' in general) is equivalent to the use of the Greek infinitive of "*esti*" as a noun (*to einai*), and also to a certain use of the participle *to on* (when used to denote the feature *to be*), and to the abstract noun *ousia*. (The word "*ousia*" also has different usages; I won't expand on this in my paper.)

My following comments in this section will focus on the notion of 'being' as a count noun. Yet this notion is closely connected with the gerund sense of 'being', since something is '*a* being' (*on*) in virtue of having 'being' (*einai*).

3.2 Positional and predicational connotations

Calling something 'a being' (*on*), frequently has an existential connotation, or more precisely a connotation of *real* existence. By calling it an *on*, the thing in question is acknowledged as a part of reality. I will call this the *positional connotation* of the term 'being'.

I also mentioned the predicational implication of the one-place ascription of being: 'To be' entails 'to be something or other' [a is $\rightarrow(\exists t)$(a is t)]. Thus the term 'being' also connotes predicational content. (A specific type of predicational content is a thing's *essence*). A being, qua being, is always *determined* in some way that can, in principle, be articulated through predication, typically involving a copula. (Aristotle emphasizes that every predicative sentence can, in principle, be reformulated so as to contain a copula.)

The talk of beings as bearers of predicational content is not necessarily equivalent to the talk of beings in the sense of positional being. All positional beings are also beings in the predicational sense, yet the reverse does not hold if your ontology allows for non-real objects. For instance, if your ontology allows for fictitious objects, such objects are beings in the predicational sense yet not beings in the positional sense.

3.3 Predicational content and the alethic aspect

As we have just pointed out, beings, in the Greek understanding, are things with a predicational content that can be expressed in statements typically involving a copula. The copula also carries a veridical implication (cf. 2.3.3) indicating that such beings are things of which something *true* can be asserted.

This is a formal property of beings qua beings, and I am calling it the *alethic aspect.* It corresponds to the *alethic connotation* of the term "being".—The alethic aspect is a formal relational property of the objects that are beings. It consists in the fact that any such object can become the referent of a true assertion. (In scholastic philosophy, this becomes the so-called transcendental sense of 'truth', according to which any object whatsoever is a possible referent of cognition or true belief.)

The alethic connotation of 'being' also introduces a contrast with mere appearance.

3.4 Predicational content and per se knowability

In the Greek philosophy of being it is generally assumed that whenever there is some instance of veridical being (i.e. some obtaining state of affairs), or some object characterized by some predicational content, this is also something that can become an object or content of *knowledge*—at least in principle. An object with its predicational content may not be (fully) knowable *to us,* due of the limits of our human cognitive capacities, but it would still be knowable *per se*, since nothing in the object itself excludes it from the range of knowable entities, given that is has a determinate predicational content.—The object is knowable *per se* in virtue of exhibiting certain characteristics.

At first blush, one would expect that *everything* has to be knowable *per se*, since everything is, in some way or other, a bearer of predicational content that can be truly asserted of it. So it comes as a surprise to learn that in Plato, for instance, everyday objects seem to be excluded from the realm of the objects of genuine knowledge (*episteme*). Yet a closer look reveals that this contention in Plato does not contravene but rather confirms the principle that every object, to the extent that it is a bearer of predicational content, is also a knowable *per se.* For Plato's reason for defending this restriction of knowability *per se* is based on the assumption that particular empirical objects, in many respects, lack fully determined predicational content. To the extent that their predicational being is not fully determined, statements or beliefs that characterize some such object cannot be unqualifiedly true; and based on the assumption that a true belief can be knowledge only if its content is unqualifiedly true, such beliefs have to be relegated to a lower class of cognitive achievement.

3.5 Temporality and the alethic aspect

Since "*esti*" is grammatically a verb, it includes temporal indices that relate to McTaggart's A-series. Ancient Greek philosophy seems to have been committed quite generally to the position of temporalism in truth theory. It was commonly held that the truth value of statements about changeable and perishable objects can vary over time: The sentence stating that George is sober becomes false as often as he gets drunk.

Now, I also mentioned that according to Plato, the lack of perfect predicational determination disqualifies an object as a member of the class of knowable entities. It is debatable whether temporality is one of the disqualifying features, or even the crucial such feature. If so, why? There are two possible explanations. The claim might be that it is not fully true to ascribe being-F to some object if it is F only for a certain time and then ceases to be F. Or it might be the stronger claim that all objects subject to time are thoroughly process-like such that none of them is F during any period without being in the process of already changing into something which is not-F[18]—which would provide an even stronger reason for holding that predicating F of such an object isn't unqualifiedly true.

3.6 Degrees of reality

It seems obvious that positional being cannot come in degrees. An object is either part of reality, or it is not (and hence merely fictitious). With respect to predicational being, on the other hand, it certainly makes sense to introduce the idea of gradation. Some object can instantiate a quality more or less perfectly. For instance, a circular shape can be exemplified by material bodies in a more or less perfect way, and in each such case, *being circular* is true to varying degrees, depending on how well the object satisfies the notion of circularity. One could also say that in a very accurately drawn circle, this property has been realized to a higher degree than in an imprecise circle. Likewise, one could say that the capacity of sight can be realized in living beings to varying degrees, both compared to other spe-

18 One might express this in the following way: For every object x susceptible to undergoing change, and for any predicative attribute φ, and for any time t during which x appears to be φ, x never IS φ, since x is in the process of becoming something different than φ during t.

cimens of the same species, and also across species. Now, it seems that the intermediate degrees in the realization of a property can themselves be viewed as distinct properties. Thus dim sight would be an intermediate capacity between accurate sight and full blindness. Hence it also seems that when such an intermediate property is realized in an object, it comes with a certain *degree* of reality that is located on a scale defined by a property at the maximum end of the scale (like perfectly accurate sight, perfect circularity, etc.).

So far, the notion of degrees of reality applies only to properties and only relative to a specific scale that corresponds to some specific property and its approximations. As yet, nothing suggests that *objects* could manifest varying degrees of reality. A specific tree, say a redwood, may be a less typical, less perfect specimen of its kind, but as a particular object, it is no less real than the more perfect specimen standing next to it. Yet the metaphysical tradition also knows the application of the concept of degrees of reality to objects. In this sense, Plato speaks quite generally of objects with a higher or lower degree of being (*mallon/hêtton onta*). How we locate objects on a scale of degrees of reality depends on what counts as the decisive *general criterion for being real.* For instance, if to be knowable is the leading criterion (as in Plato's middle-period epistemology and ontology), then objects that are more knowable than others would rank higher on this scale of reality. If, on the other hand, causal capacity is the criterion (a criterion first discussed in one of the later dialogues of Plato, the *Sophist*), then a more powerful being whose causal capacity is richer or more powerful could count as having a higher degree of reality than less powerful beings.

4. Parmenides' Strategy

In the history of ancient philosophy, Parmenides was the first to pursue the idea that the analysis of the notion of being and its logical relations could be essential for an adequate philosophical account of reality. He even went so far as to claim that everything could be reduced to this very notion of being.

What was his justification for singling out the notion of being as the starting point of philosophical inquiry? Parmenides claims that the philosophical investigation has to begin by distinguishing and making a choice between two basic assumptions. The one assumes being ("It IS and cannot not be"), the other not-being ("It is not and needs must not be"). When he introduces these alternatives (B2 *Diels-Kranz*, cf. B8.15-6), he does not tell us explicitly why this is the choice to be made at the beginning of the investigation, and so we have to guess what his reasons are. This is complicated by the fact that the "is" in Greek can have different meanings or connotations, as we have seen.

Since we are dealing here with a 1-place use of being, the generally recognized possibilities are the existential *esti*, the veridical *esti*, and the elliptical use of the copula (cf. 2.3). Now, I think that the initial argument is easiest to interpret if we construe the *esti* in the sense of existence, or more precisely, of positional being.—Ancient philosophers understood their enterprise as a scientific investigation of certain structures and features of reality, and ultimately of reality in its totality. Thus their subject-matter could always be described as some being, or the totality of beings, in the positional sense of 'to be' (i.e. as denoting objects which are not merely constructions of our mind, or of our linguistic conventions, but exist independently of us). It is therefore not at all implausible that Parmenides should have focused on this aspect of the notion of being.

On the other hand, it is also characteristic of any kind of investigation that it aims at knowledge and tries to avoid mere appearance. It wants to find out the real properties of its objects, not some merely apparent ones. Thus every kind of investigation also relates to being under its alethic aspect (cf. 3.3). It is a very conspicuous feature of Parmenides' text that it juxtaposes, and contrasts, a truthful account of reality with a seemingly plausible but 'deceitful' account, the latter being based on the common-sensical but false assumption of the possibility of change and of mixture between opposites (B1.28-32, B2.4, B8.50-61). Since he, moreover, also links the two concepts of truth and being (B8.18), it is obvious that the alethic contrast between truth and mere appearance is very relevant for his conception of being.

Yet from his point of view, the positional and the alethic aspects in the concept of being are probably inseparable, since the object of inquiry, together with its characteristics, would not be an instance of knowable truth if it were not a *real* object. If our thought (*noein*) gets a hold on the truth, it grasps and represents a real object such as it really is. (We can connect this with the types of 'veridical unfolding' described in section 2.3.3.) This might be the background to his claim that any thought whatsoever is a thought "that it is" (B8.34)—because, to paraphrase it in our own words, a thought always tries to represent (some aspect of) reality.[19]

So far we have seen why Parmenides can take it for granted that any kind of systematic inquiry has being or beings as its objects. Yet this observation alone would not suffice for justifying the claim that the notion

19 The grammatical structure, though, of 8.34 is very uncertain, which makes it impossible to reach an uncontroversial interpretation of this line, which is deplorable since it articulates a very important assumption in Parmenides' argument.

of being should be *central* for philosophical inquiry. There were other notions available, like *cosmos* or *physis*, that could serve as names for the subject-matter of philosophy-science in its entirety. Moreover, since every real feature of the world can be said *to be*—both in the veridical and in the positional sense, the characterization of the object of investigation as being does not yet seem to entail anything about how reality is structured.

So why single out the notion of being although it is, apparently, devoid of any descriptive content?

I submit that the crucial idea of Parmenides is that the ascription of being is not only a priori true of any object of inquiry, but that 'to be' is itself a basic descriptive predicate that entails further descriptive predicates due to its opposition to not-being. More precisely, it entails that all those predicates have to be dismissed which presuppose the possibility of not-being *simpliciter*, and that their polar opposites, *if those do not involve not-being*, are necessary attributes of being. (Note that Parmenides' text nowhere claims the impossibility of negative predication or 2-place not-being. If it did, his attempt at listing certain characteristics of that-which-is would encounter insurmountable difficulties since every characterization also entails a negation. Yet there is no need to impute such a view to him.)

Moreover, he thought that the a priori characterization of the object of investigation that could be derived from the predicate 'to be', would establish that our ordinary experience of the world, and our linguistic means of representing the world, are false and based on self-deception. The aim of refuting the truth of our ordinary experience of the world with the help of some powerful a priori reasoning obviously provided the main motive for his approach.

As I see it, it is the fatal *prôton pseudos* of his approach that he assumes not only that being *simpliciter* holds a priori of the object of investigation but also that being *simpliciter* is a descriptive first-order predicate which, due to its opposition to 1-place not-being, entails further characteristics. If being *simpliciter* is a first-order predicate at all, it is certainly only a formal predicate that does not entail any specific descriptive content (which is the meaning of Kant's dictum, frequently misunderstood, that 'to be', or 'to exist', is not a 'real' predicate).[20] To see whether or not my claim is justified, we need to look a bit more closely at how the characteristics of being are derived from the bare antithesis of being and not-being.

The derivation of the characteristics of being takes the following crucial steps:

20 Cf. L. Honnefelder, *Zeit und Existenz*, in H. M. Baumgartner (ed.), *Das Rätsel der Zeit*, Freiburg/München 1993, 332-362, 337.

First, the ascription of being is secured by pointing out that the object of investigation cannot be conceived as not being since what-is-not is inconceivable (B2.6-8 with B.3, cf. B6.1-2, B8.8-9, 17, 36-7).

Secondly, it is pointed out that there is also no middle ground, since being and not being exclude each other (B6.4-9[21]). It either is or it is not, *tertium non datur.*—This strict exclusion is probably also the reason for the modal qualifications that he adds in B2, lines 3 and 5: If it is, it *cannot* not be since it strictly excludes not-being, and if it were not, then necessarily so, since not-being could not adopt being. Furthermore, this strict exclusion is also Parmenides' reason for rejecting the idea that being could come in degrees (B8.23-4, 33, 47-8). There is no more or less of being—presumably because this gradation would presuppose that being and not-being can somehow mix, or that being can gradually approximate not-being.

Next, it is shown that the exclusion of not-being from being entails a number of more specific characteristics of being (B8): that it excludes coming-to-be and perishing; that this being is not subject to the lapse of time; that it is not divided into parts since it is everywhere equally 'full' and hence perfectly continuous; that it is free of any lack and hence is a complete whole, and that therefore there is no reason why it should move in any direction or change its appearance (color); that since it is a complete whole that does not lack anything, there cannot be anything outside of, or separate from, being; and that as a complete whole it has limit, and hence a shape, which is described as spherical in view of the fact that only this shape allows that there be an equal amount of being from the centerpoint in every direction. The outcome, hence, is that that-which-is is a unique object which is everywhere equally full or dense, has a perfect spherical shape, and enjoys some form of eternity. (Note that his suggestion that thought cannot be separated from being [B8.34-6, B3, B4] does not have to mean that thought and being are simply identical. To deny the separability of thought from being is not the same as asserting their identity.)

The interpretation of the details of the argument is extremely controversial, due to the obscurity of many of Parmenides' grammatical constructions and the vagueness of many of his argumentative moves. Thus, already this short summary of his conclusions contains some very controversial claims. I can't engage in this discussion here and will restrict myself to giving a brief outline of how I think the results regarding the unity, the uniqueness, and the non-temporal character of being are established based on the rejection of not-being.

21 Perhaps also B7, but it is a moot point whether this fragment refers to the second or the third way.

Let's begin by looking at the reason he gives for the first crucial step in the argument, the rejection of the possibility of not-being.

The rejection of not-being is based on the claim that something which is not would be inconceivable, the apparent reason for this claim being that it would be simply *nothing*, and one cannot conceive of something which is nothing. This latter claim is plausible enough— something which is nothing would be a pseudo-object, but how does he get from "is not" to "is nothing (at all)"? The move seems to presuppose that "not: x is" entails "(φ) (not: φx)" [with φ as variable for some predicate], which is the contrapositive to what I called the predicational implication of existential *esti* in section 2.3.1. The easiest way to justify this entailment is to say that for an object to have or instantiate certain properties it needs to exist. This has been ridiculed as the 'hatstand model of predication'[22], but it is not at all an absurd philosophical position. Note again that this entailment is not to be confused with the absurd claim that negative predication entails non-existence (For any predicate φ, if x is not φ, x is not *simpliciter*).

Since the remainder of Parmenides' argument, I think, presupposes that 'to be' and 'not to be' are treated as first-order-predicates, I will interpret them along the lines of what I have called the predicate of real existence or positional being.

Although Parmenides' argument uses 'to be' and 'not to be' as first-order-predicates, 'to be' is not treated as a property that inheres in something else. The argument does not leave room for the kind of distinction between that-which-is (participle sense) and its being (gerund sense) that would be analogous to the distinction between an object and one of its properties. The term 'being' is, rather, used like a certain kind of mass-noun—denoting a substantive feature that can somehow fill space (as water or light can fill space). In this specific respect, 'being', as construed by Parmenides, is similar to the primary 'stuffs' which the philosophers in the Ionian tradition before Parmenides talked about. At the same time it retains its existential or 'positional' meaning, and it is this construal of existence as a descriptive predicate what I take to be the main logical failure of his approach.

How does he get from the rejection of not-being *simpliciter* to the affirmation of the unity, uniqueness, and non-temporality of being?

The fact that several times in the course of his deduction of the characteristics of being, he reiterates both the necessity of a choice between the assumption of being and not-being, and the inconceivability of being, makes it quite obvious that he views this as the crucial foundation to his

22 Cf. G.E.L. Owen, *Plato on Not-Being*, in *Logic, Science, and Dialectic. Collected Papers in Greek Philosophy*, London 1986, 105-137, 123.

argument. The main idea behind his deduction shows up quite clearly in his first argument against the possibility of coming-to-be. He claims that for something to come to be, it would have to develop out of something which is not, or change from what is not into something which is (B8.6-8). (The Greek formulations use the metaphor of growing: growing from what is not, or is nothing: "*auxêthen ... ek mê eontos*", "*tou mêdenos arxamenon phyn*", B8.7, 10.) Thus it would have to be conceivable that it is not (or was not) at a certain time. Yet since not-being *simpliciter* is inconceivable (B8.9-10), the notion of coming to be has to be rejected as well, as the latter notion presupposes the former one.

It is not clear whether or not he also offers an argument against the possibility of perishing or simply states that the inconceivability of not-being also entails the inconceivability of perishing. Of course he could offer a similar argument which points out that to perish means to change from a state of being into a state of not-being. I am inclined to follow Barnes in suspecting that lines 12 *f.* actually serve this purpose. But this can be no more than an uncertain conjecture since the Greek here is very obscure.[23] Most interpreters construe this line differently and assume that the argument against perishing is left implicit.

Parmenides then goes on to prove the impossibility of past being and future being as expressed by the past and future tenses of the verbs for being or coming-to-be (B8.19-20, cf. 5). Some of the Greek is again quite obscure (especially lines 19-20), and the content of the argument anything but clear. Yet at least one can say with some confidence that he links the notion of past and future coming-to-be with the idea of non-existence in the present (following a construal of the lines 19-20 that is supported by Kahn and Barnes[24]). One problem with this move is that it seems an obvious non-sequitur to claim that something which came to be in the past, is not (now). Perhaps his idea is that if something has an origin in the past or will come to be in the future, it cannot 'be', strictly speaking, since what is, necessarily is, due to the impossibility of not-being.

Since he seems to endorse the notion of the present (or "now", cf. B8.5), and a few lines later even speaks about being as "remaining" the same and in the same location (29-30), the underlying idea is most plausibly interpreted as that of a 'standing now', which is one way of conceiving

23 Cf. J. Barnes, *The Presocratic Philosophers*, London/New York 1982, 188-90; text based on Reinhardt's emendation for line B8.12.

24 Cf. Ch. Kahn, Review of "L. Tarán, *Parmenides: Text, Transl., Comm. etc.*", Gnomon 40 (1968), 123-133, 128 f.; Barnes, *The Presocratic ...* [fn. 23], 190 f.

eternity.[25] Being is hence non-temporal only in the sense that it is not subject to the *lapse* of time.

But why would he reject the idea that being has a temporal career that includes a past and a future, if he has only shown that it does not have a temporal origin or end? The classical Greek theories of time view time as an aspect of motion and change. It is therefore plausible to assume that the intuition behind Parmenides' claim is an understanding of temporal qualifications according to which the talk of a past and a future is meaningful only against a background of change: Since being cannot originate from something else or change into something other than being, it is also not subject to the lapse of time, given that such temporality presupposes the possibility of changing into something that it wasn't before. (It is noteworthy and somewhat puzzling for modern commentators that Parmenides' argument against the possibility of locomotion comes in a later passage, B8.29-33, and is not used, apparently, for his case against past and future. This is also a problem in the ancient perspective, since for the ancients it was the movement of the celestial bodies that provided the measurement of time.)

Whatever the exact import and merits of this argument against the temporality of being, the crucial result for my purpose here is that the argument is based on the rejection of coming-to-be and perishing, which in turn is based on the exclusion of not-being from that which is.

The argument for unity has, in principle, the same foundation (B8.22-25). The unity of that-which-is is a consequence of the rejection of its having parts. The argument for this conclusion (or at least the clearest part of it) establishes the equal degree of 'fullness' in being in order to infer its perfect continuity, which in turn is the basis for claiming that being is not divided into parts. Now, equal fullness is a consequence of the rejection of gaps or unequal degrees of fullness within being (see also B8.46-48). The rejection of gaps in being is most plausibly understood as a consequence of the inconceivability of positional not-being, since gaps within being would be nothing else than limited zones of not-being. The reason why degrees of being are inconceivable is not explicitly stated. Yet this is again most plausibly connected with his idea of a strict alternative of either being or not-being, since this restriction to a duality of exclusive alternatives does not seem to leave room for intermediate levels. Being is either 'completely full', or it is not at all.

We find this type of argument explicitly stated in a later passage that tries to establish the more general point that being can't exhibit any lack

25 Cf. G.E.L. Owen, *Plato and Parmenides on the Timeless Present*, in *Logic, Science, and Dialectic. Collected Papers in Greek Philosophy*, London 1986, 27-44.

or deficiency because "if it lacked anything, it would lack everything" (B8.33). This is of course incompatible with the idea of degrees of reality discussed above, in section 3.6: Parmenides rejects the idea that there can be a diminished degree of being in view of his initial statement of a strict alternative between being and not-being, which also entails the rejection of any kind of intermediate position that would somehow combine being and not-being (the so-called 'third way'). A lack can be viewed as involving the non-existence of some relevant part, and would thus fall under Parmenides' strictures against positional not-being, while a diminished degree of being would be viewed by him, I assume, as erroneously presupposing the possibility of some sort of mixture between being and not-being.

Note how the claim that "in its entirety, it is full of being", without more being here or less being there (B8.23-24), clearly indicates Parmenides' interpretation of the notion of being as a feature which somehow "fills". In section 2.3.1, I talked about the predicational implication of positional being. Given the way this implication is defended by Plato or Aristotle, it means that being can always be specified in reference to some specific descriptive content, some specific 'essence'. Thus being is not itself a descriptive predicate but rather requires a descriptive ('categorial') predicate as its supplement since nothing simply exists, but exists only by way of exhibiting some specific essence other than being *simpliciter* (with the sole exception of the general Form of being which Plato's *Sophist* and *Timaeus* talk about—a conception rejected by Aristotle). In Parmenides, we don't find this predicational implication, rather the predicate of positional being itself is treated as a fully specific descriptive predicate that names a space-filling feature. This is also the conceptual basis for the claim of the homogeneity of being (B8.22, see also 49): Only if we assume that being *simpliciter* does not need to be supplemented by some other predicative content but is in itself a fully sufficient, fully determinate characterization of an object, does the equal degree of being entail its homogeneity. Otherwise there could be different zones in being with different specifications that all have the same degree of reality.

The argument just discussed also seems to serve as a foundation for his assertion of the singularity or uniqueness of being: Being (that which is) is not only a unity with no parts, there also can't be anything outside, or apart from, being (B8.36-38), which establishes the sort of monism that was commonly attributed to Parmenides in antiquity and which seems to me still the right interpretation of Parmenides (notwithstanding the fact that some scholars now argue against this monistic interpretation of Parmenides). The uniqueness of being is inferred from the fact that being is a complete whole and not subject to locomotion or change (B8.38). Now, the details of this inference remain again rather obscure. But at least one aspect of its justification seems to be that being could not be a complete and continuous 'whole' if something could *exist* apart from being. For, if this were the case, we would have some being outside being, but this cannot be true, given that being in its entirety is a continuous whole, as has been proved on the basis of the rejection of not-being (B8.30-33). (The idea that there could be empty space or void beyond the limits of being is of course also rejected by Parmenides—at least implicitly—since being 'fills', and absence of being is impossible.)

5. Concluding Remarks

The main result of our brief survey of Parmenides' strategy is that he takes up the notion of being in its existential (positional) meaning but uses it at the same time as if it were a descriptive predicate that designated a fully determinate space-filling feature. The veridical connotation is also present, i.e. that-which-is is also an instance (and in fact the only instance) of knowable reality. Yet the crucial, and ill-conceived, move is the one that conflates the positional with a descriptive predicate.

The result that he construes 'to be' as a fully specified descriptive predicate, can be gathered from the fact that 'to be', in Parmenides' usage, does not require the sort of supplementation that we know from Plato and Aristotle and which we dubbed the predicational implication: The difference comes out most clearly in Aristotle who completely and unequivocally rejects the idea that the word 'to be' as such has any specific descriptive content and emphasizes that the assertion that some x *is* (exists), always warrants the question *what* it is. The answer to this question would have to provide some genuine predicate other than "is". In other words, in Aristotle's perspective something can BE (or exist) only by way of instantiating some specific descriptive content or 'essence', while in Parmenides' perspective 'to be' is already a full specification, or, to put it in Aristotelian terminology, a fully specified essence.

To put Parmenides' notion of being in context, we can first point to the Ionian conception of an original and originative stuff. The Milesians postulated some primary kind of stuff out of which everything else originates. The notion of this stuff (unlike Aristotle's conception of matter) meant to designate something fully determinate, fully specified, capable of subsisting by itself and filling space: water in Thales, the 'boundless' in Anaximander, *aer* (mist) in Anaximenes. It was, moreover, invested with the attributes of a rational, governing principle.

In Heraclitus, fire is identified as the stuff that embodies the governing rational principle of the cosmos, and it is said of fire that it "always was and is and will be" (22B30). Likewise it is said of the governing rational principle as such (the *logos*) that it always is (22B1). Yet at the same time it is affirmed of fire, which embodies this *logos*, that it dies and that its death is the life of some other, opposing, stuff (water).[26] He sees the death of fire as part of a cycle in which fire will always, eventually, be born again through the death of some other stuff (cf. 22B30, B31, and B76). Thus Heraclitus paradoxically maintains that fire is *continuously* preserved as the governing principle through a cyclical process in which it ceases to be and then comes to be again. Thus this primordial and originative stuff both is and is not, or is preserved and perishes—a combination of contradictory characterizations characteristic of Heraclitus' understanding of the kind of 'logic' that governs the cosmos.

Parmenides' notion of being takes up the positional and veridical connotations present in Heraclitus' assertion that the governing *logos* of the cosmos always is. It also assimilates this notion to the way in which the Ionians talked about the basic feature of reality as a space-filling stuff.[27] Yet he rejects the Ionian notion that the basic feature can change into something other than itself, and also Heraclitus' idea that being is inseparable from not-being (cf. 28B6.4-9). He uses his insistence on the principle of non-contradiction, together with his argument about the inconceivability of not-being, to construct an argument that exposes our perception of plurality and change as mere appearance. A consequence of his ill-conceived 'logic of being' is the rejection of distinct tenses (McTaggart's A-series), against the earlier philosophers who describe reality as temporally structured and 'tensed' (witness the only extant fragment of Anaximander (12B1) and Heraclitus' fragment 22B30).

The Greek philosophy of being begins, hence, with an argument that tries to expose the lapse of time and the distinction of tenses as mere appearance without truth. The subsequent theories that use the notion of being without adopting the paradoxical Eleatic view-point, most interestingly those of Democritus, Plato, and Aristotle, can be viewed as various ways of reconciling the notion of being with the manifest reality of change and temporality.

Democritus uses the term 'being' as a label for the atoms, and the terms 'not-being' and 'nothing' as labels for the void (67A6a, 68A37).

26 The details of Heraclitus' conception of the cycle of basic stuffs are hard to reconstruct and don't need to concern us here.

27 When I use the term 'space-filling', I don't assume that the early Ionian thinkers already had an abstract, theoretical, notion of space. I am only appealing to a pre-theoretical notion of space as something that can be *filled* with objects or stuffs.

Atoms and the void are the two basic elements of his metaphysics. In using these Parmenidean labels, he also intentionally takes up some of the formal characteristics of Parmenidean being: Each basic being, or atom, is an instance of the fullness of being, has shape but no parts (indivisibility), and is indestructible and not subject to alteration. Not-being or nothing, on the other hand, is equated with that which is not filled at all, i.e. with complete absence of being, or *void* (based on the Parmenidean view-point that being somehow 'fills'); yet void is allowed to be a part of the cosmos, and this, in turn, is thought to provide the basis for the existence of a plurality of 'beings', separated by the void, and for motion, which is motion through the void. On the back of the concept of motion we also admit temporality. Macroscopic objects, moreover, can come to be and perish since they are nothing other than the results of atoms combining and recombining into complex compounds.

Plato's ontology of Forms also borrows from Parmenides' conception of being, as is well known, yet avoids the disastrous consequences of Parmenides' ontology because it does not construe being as a first-order descriptive predicate but rather as a kind of formal, trans-categorial, predicate that can be predicated of objects and of Forms alike and can serve to demarcate the ontological distinctions that hold between empirical objects and intellectual Forms. Descriptive contents like 'circle' or 'justice/just' can, according to Plato's middle-period ontology, function as objects of our perception and cognition in different ontological modes. The primary ontological distinction is between such a content as the 'thing itself', which is an intellectual, non-temporal, unambiguously determined object (i.e. the Form, which is nothing other than a reified descriptive content), and the transient, often imperfect or context-dependent, reflections of the 'thing itself' in perceptible objects. 'Being' in the strict sense applies to Forms, with some of the formal characteristics of Parmenidean being, namely lack of temporality, strict exclusion of the opposite, and hence unambiguous and unchanging determinateness (e.g. Phd. 74b-c, 78d-e, Symp. 210e-211b, Rep. 478e-479e, 508d, Phil. 59a-d). In a laxer sense, the empirical objects can also be called beings, yet they are viewed as having a 'lesser degree' of being (cf. above, 3.6)[28]; in other words, they are supposed to be less real, and this idea of a varying degree of being is again

28 This view is implied in calling the Forms *mallon onta* (cf. Rep. 479c-d, 515d, 585b-e) or *ontôs/teleôs onta* (cf. Rep. 597A5, Phdr. 247c-e, 249c, Tim. 28a, 52c5-6, Soph. 248a11, Phil. 58a, 59d). On the general ontological and epistemological background cf. G. Vlastos, *Degrees of Reality in Plato*, in R. Bambrough (ed.), *New Essays in Plato and Aristotle*, London 1965, 1-19; Szaif, *Platons Begriff...* [fn.4], 72-152, 183-324; M.F. Burnyeat, "Plato on Why Mathematics is Good for the Soul," in T. Smiley (ed.), *Mathematics and Necessity in the History of Philosophy* (=*Proceedings of the British Academy* 103), New York/ Oxford, 1-81.

tied to the role of the notion of being as a formal concept which is not itself descriptive but rather qualifies the ontological mode in which descriptive contents can be present. As a second-order feature of Forms, the concept of being does not conflict with the idea that there are many diverse descriptive contents or Forms.

Moreover, Plato introduces a concept of participation, or exemplification, that allows for descriptive features to be present in, or reproduced by, many other objects, without destroying the unity and completeness of this feature as such. This notion of participation can then also be applied to the *Form of being* as such, which means that other Forms (and to a lesser degree also concrete objects) can participate in it (and thus 'be') without diminishing or compromising the unity and non-temporal subsistence of the *Form of being* as such. Thus it is guaranteed that the stipulation of a Form of being will not lead to the disastrous Parmenidean consequence of reducing everything other than *being itself* to non-being.

Although Plato's Forms share the non-temporal character of Parmenidean being, he also allows for a temporal reality since participation in Forms (i.e. the instantiation of Forms in the perceptible realm) is temporal. Yet neither he, nor Democritus before him, can provide a fully satisfactory theoretical account for the temporal persistence of objects that exist for a limited span of time and can undergo change without thereby ceasing to be, since such a theory would require an exploration of a certain type of predicates, predicates that designate 'substantial forms' (to use the Aristotelian terminology), and we don't get this from either Democritus or Plato. Plato, to be sure, recognizes that it is not sufficient to characterize the perceptible realm as a domain of objects that are always in the process of changing. For an object to continue through time, it also has to exhibit some characteristic that does not change. Total flux would reduce all empirical objects to mere momentary entities (cf. Tht. 156a-157c, 181c-183b, Crat. 439c-440b). It is also not sufficient to introduce souls as the basis for the continuation through time of individual entities, since there are also individual entities with well-defined numerical identity that persist through time although they don't have a soul, especially artifacts and social institutions. We get hints in various contexts that the production of some new particular artifact or institution, which 'is' once it has come to be (*genesis eis ousian*, Phil. 26d8), depends on the realization and preservation of some unified functional structure or proportion in some material (e.g. Gorg. 503e, Polit. 283c-284d , Phil. 25b-27b). Aristotle's theory of persistent objects will elaborate on this starting-point by exploring the ontological foundations of predicates that signify a structural unity, or substantial form, which subsists in a material substratum and thus constitutes an object that is subject to qualitative and quantitative change and locomotion, yet continues to exist while changing.

The task of a theory of persistence for objects that are subject to change is to identify a descriptive content which does not change during the course of the object's temporal career, and provides the criterion for determining both the beginning and the end of the object's temporal existence and, furthermore, for distinguishing what is a part and what is not a part of this object at any moment in its career. The main target of his theory of substantial forms is the phenomenon of persistence in the domain of living things. Persistence, here, can be equated with continued living, i.e. with the continuation of a specific form of life in a material substratum (whose elementary components can continually be exchanged through metabolism). Now, an individual living being grows and then turns to decaying in the course of its temporal career, and thus it continuously changes the way in which it performs its characteristic life activities. The challenge, therefore, for the Aristotelian theory of substantial forms of living things is to identify what is unchanging during the entire career of such a being. It is not enough to say that what is unchanged in a squirrel's life-span is its *being a squirrel*, since this answer would be question-begging. Rather, the analysis of the substantial form of a squirrel would have to explain what it is *to be a squirrel*, and to do so in such a way as to give us an informative answer regarding what is unchanging during the life-span of each being of that kind.

It is not the task of this paper to pursue this question further. Suffice it to say that for Aristotle the paradigmatic case of being is the temporal existence of a living thing which preserves a certain substantial form throughout a dynamic, goal-directed process of growth and reproduction. This notion of being preserves the idea that being requires the unchanging presence of some specific descriptive content, yet at the same time it links this content, this form, with a potential for growth and reproduction, which are essentially temporal notions. We have thus gone a long way from Parmenides, who rejected as mere appearance the idea of life, with its cycle of "hateful births" (B12.4) and inevitable deaths, and tried to justify a notion of being that excluded the lapse of time.

References

Barnes, J. (1982) *The Presocratic Philosophers*, London/New York.

Brown, L. (1986) 'Being in the *Sophist*: A Syntactical Enquiry', *Oxford Studies in Ancient Philosophy* 4, 49-70.

Brown, L. (1994) 'The Verb 'to be' in Ancient Greek Philosophy' in Everson, S. (ed.), *Language (=Companions to Ancient Thought, vol. 3)*, Cambridge, 212-236.

Burnyeat, M.F. 'Plato on Why Mathematics is Good for the Soul' in: Smiley, T. (ed.), *Mathematics and Necessity in the History of Philosophy* (=*Proceedings of the British Academy* 103), New York/ Oxford, 1-81.

Carnap, R. (1950), 'Empiricism, Semantics, and Ontology', *Revue Internationale de Philosophie 4*, 20-40.

Frege, G. (1918/19) 'Die Verneinung. Eine logische Untersuchung', *Beiträge zur Philosophie des deutschen Idealismus* 1, 143-157.

Honnefelder, L. (1993) 'Zeit und Existenz' in: Baumgartner, H. M. (ed.), *Das Rätsel der Zeit*, Freiburg/München, 332-362.

Kahn, Ch. (1968), Review of "L. Tarán, *Parmenides: Text, Transl., Comm. etc.*", Gnomon 40, 123-133.

Kahn, Ch. (1973a) *The Verb 'Be' in Ancient Greek* (= W. M. Verhaar [ed.], *The Verb 'Be' and its Synonyms, part 6*), Dordrecht.

Kahn, Ch. (1973b) 'On the Theory of the Verb "To Be"' in: Munitz, M. K. (ed.), *Logic and Ontology*, New York, 1-20.

Kahn, Ch. (1981) 'Some Philosophical Uses of "to be" in Plato', *Phronesis* 26, 105-134.

Kahn, Ch. (2004) 'A Return to the Theory of the Verb be and the Concept of Being', *Ancient Philosophy* 24, 381-405.

Kremer, K. / Wolf, U. (1984) 'Ontologie' in: *HWPh*, vol. 6, Basel, 1189 ff.

Leigh, F. (2008) 'The Copula and Semantic Continuity in Plato's Sophist' in: *Oxford Studies in Ancient Philosophy* 35.

Lyons, J. (1968) *Introduction to Theoretical Linguistics*, Cambridge.

Owen, G.E.L. (1986a) 'Plato on Not-Being' in: *Logic, Science, and Dialectic. Collected Papers in Greek Philosophy*, London, 105-137.

Owen, G.E.L. (1986b) 'Plato and Parmenides on the Timeless Present' in: *Logic, Science, and Dialectic. Collected Papers in Greek Philosophy*, London, 27-44.

Quine, W. V. O. (1980) 'On what there is' in: *From a Logical Point of View*, revised ed., Cambridge (Mass.), 1-19.

Reiche, M. (2006) 'Nonexistent Objects' in: *Stanford Encyclopedia of Philosophy*, http://plato.stanford.edu/entries/nonexistent-objects/

Szaif, J. (1996) *Platons Begriff der Wahrheit*, Freiburg/München.

Szaif, J. (2003) *Der Sinn von „sein". Grundlinien einer Rekonstruktion des philosophischen Begriffs des Seienden*, Freiburg/München.

Szaif, J. (2007) 'Plädoyer für eine formalsemantische Rekonstruktion des philosophischen Begriffs des Seienden' in: Lutz-Bachmann, M., Schmidt, Th. M. (eds.), *Metaphysik heute — Problem und Perspektiven der Ontologie / Metaphysics Today — Problems and Prospects of Ontology*, Freiburg/München, 146-195.

Tugendhat, E. (1992) 'Der Wahrheitsbegriff bei Aristoteles' in: idem, *Philosophische Aufsätze*, Frankfurt a. M..

Vlastos, G. (1965) 'Degrees of Reality in Plato' in Bambrough, R. (ed.), *New Essays in Plato and Aristotle*, London, 1-19.

Williams, C. J. F. (1981) *What is Existence?*, Oxford.

NEW IDEAS ON SUBJECT AND IDENTITY IN MEDIEVAL LOGIC

SIMO KNUUTTILA

Summary

Many medieval logicians regarded the copula as indicating the numerical identity of things signified by the terms. This identity theory of predication, which influenced Leibniz's ideas similar to later theories of logical quantifiers, was found particularly useful in analysing various problems of Trinitarian formulations. Numerical identity was treated as symmetric and reflexive but not necessarily transitive in the Trinity, which led to detailed discussions of the modes of identity and related questions partially analogous to those in the contemporary material constitution debate. An interesting part of the new predication theory was the analysis of the identity of subjects in tensed and modal propositions.

In describing the logical form of propositions, Leibniz writes:

> When I say that every A is B, I mean that anyone of those which are called A is the same as some of those which are called B, and this proposition is called the universal affirmative. When I say that some A is B, I mean that some of those which are called A is the same as some of those which are called B and this proposition is called the particular affirmative.[1]

Some historians have seen in these and related formulations the same idea as was later spelled out in the logical theory of quantifiers.[2] In order to explain his point, Leibniz employs quantifying phrases which pertain to the individuals covered by the terms and the identificatory expression 'is the same'. This device is also used in what is called the medieval identity theory of predication – in fact Leibniz refers to scholastic theories among

1 G.W. Leibniz, *Mathesis rationis*, in Louis Couturat (ed.), *Opuscules et fragments inédits de Leibniz* (Paris,: Felix Alcan, 1903), 193.

2 Wolfgang Lenzen, *Calculus universalis. Studien zur Logik von G.W. Leibniz* (Paderborn: Mentis, 2004), 99-131.

the sources of his view.[3] Even though scholastic formulations were often put forward as explications of Aristotelian syllogistic premises, medieval writers were well aware that the new analysis, which was also employed by Leibniz, was not found in Aristotle. I think that the philosophically interesting difference is that in the new theory the copula is taken to express the numerical identity of the subjects which the terms stand for. I shall deal with medieval discussions of this theory and some related developments in theological analysis of Trinitarian statements, particularly the ideas of identity advanced in these contexts. In the last section, I add some remarks on how the idea of the identity of logical subjects was applied in the analysis of tensed and modal propositions.

1. Identity and Difference in Oxford Realists

Let us begin by taking a look at the late fourteenth- and early fifteenth-century discussions by the so-called Oxford realists who, to be sure, did not introduce the new theory of identity, but who made use of it – this is of some interest because many earlier proponents of the theory were nominalists and therefore opponents of realists. The famous Italian logician Paul of Venice studied in Oxford in 1390-3, where he became familiar with the discussion of general objects by followers of John Wyclif. Like many of these 'Oxford realists', Paul of Venice also published a treatise *On universals* which explores the logical behaviour of universal concepts and the metaphysical nature of universal entities.[4] In arguing for his realist position, which implies that the common human being is really the same as any individual human being, for example, the author first explains the traditional medieval distinction between personal (suppositional) and essential (simple) supposition; the former pertains to the terms of a sentence which refer to (supposit for) non-universal singular beings and the latter to the terms which refer to universal or common features. While some terms, such as proper names, only have a personal supposition and some others only an essential supposition, for example, the abstract terms 'humanitas' or 'animalitas', there are terms which may have a personal suppo-

3 *De lingua philosophica*, in G.W. Leibniz, *Sämtliche Schriften und Briefe*, IV.4, ed. Heinrich Schepers et al. (Berlin: Akademie-Verlag, 1999), 186 (889). Leibniz also refers to a similar analysis by Johann Raue (1610-79) in this context; see Maria Rosa Antognazza, 'The Defence of the Mysteries of the Trinity and the Incarnation: An Example of Leibniz's 'Other' Reason', *British Journal for the History of Philosophy* 9 (2001), 295-8.

4 The *Quaestio de universalibus* has been partially edited by Alessandro Conti as an appendix of Johannes Sharpe, *Quaestio super universalia*, ed. A. Conti, Unione Accademica Nazionale, Corpus philosophorum medii aevi, testi e studi 9 (Florence: Olschki, 1990), 199-207.

sition or an essential supposition; for example, 'homo' and 'animal' have a personal supposition in some propositions, an essential supposition in others, and indifferently one or the other in some propositional contexts. If the terms of syllogistic premises do not have the same sort of supposition, the terms of the conclusion should not be taken to have a definite kind of supposition, the terms of a correct conclusion being indifferent with respect to common and individual entities. For example, the premises 'All human nature is Socrates' and 'All human nature is Plato', having terms which refer to common and singular things, do not imply that Plato is Socrates; what instead is correctly derived is the conclusion: 'Plato is something that is Socrates'. The term something (*aliquid*) indifferently stands for common nature and individual singular beings. The argument then goes on as follows:

> One can solve all paralogisms in divine matters in this way, remarking that this does not follow: 'Every divinity is the Father, the Son is the divinity; therefore the Son is the Father'; what follows instead is this: 'The Son is something that the Father is'. Similarly this does not follow: 'Every Father generates, the divine essence is the Father; therefore the divine essence generates'; instead, only this follows: 'Something that is the divine essence generates', where 'something' is indifferent with respect to both suppositions.[5]

After the remarks on supposition, the distinction between identical predication (*praedicatio identica*) and formal predication (*praedicatio formalis*) is explained:

> Identical predication is that by which it is indicated that what is signified by the subject term is really the same as what is signified by the predicate, for example, 'A human being is an animal'. Generally speaking, this predication is formed by the terms of first intention when these are predicated of each other without qualification in propositions which are true or false or necessary or impossible. Formal predication, again, is that by which it is indicated that the subject and the predicate come together because of the same formal rationale, and such a predication is formed by the terms of first intention with specifications such as *formaliter*, *per se*, *in quantum*, *ut sic*, and others, for example 'A human being is *per se* an animal', that is, 'A human being according to the formal rationale is an animal.' This formal rationale of a human being is called humanity ... Further, the terms of second intension constitute a formal predication when they are predicated of each other without qualification.

As in the discussion of suppositions, the author adds that if one premise involves an identical predication and the other a formal predication, there is no inference of a conclusion with a formal predication.[6]

Following the identity theory of predication as formulated by William Ockham, John Buridan and others, Paul of Venice regards the basic af-

5 Ibid. 200-1.

6 Ibid. 201-2.

firmative proposition as an assertion in which things signified by the subject term are the same as those (or some of those) signified by the predicate term. The sameness of identical predication is the sameness of supposita or logical subjects as such, not sameness in the sense of sharing something in common as in formal predication. 'A human being is a donkey' is not equivalent to 'The same (*idem*) is a human being and a donkey' nor is 'Socrates is Plato' equivalent to 'The same is Socrates and Plato'. The former propositions, which are false, involve an identical predication, and the later ones, which are true, involve a formal predication, for 'idem' in these stands for something which is common to two things. Paul of Venice suggests that the sameness of identity predication can be explicated using the adverbial terms 'identice' and 'realiter', which do not refer to anything shared by the subjects in the way 'idem' could be taken to do. These signs of 'real identity' can be added to propositions which are intended to express identical predication.[7]

The questions of this treatise derived from the discussion which was introduced by John Wyclif and refined by several late fourteenth- and early fifteenth-century writers, such as the Englishmen Robert Alyngton, William Milverley, William Penbygull, Roger Whelpdale, John Tarteys and the German Johannes Sharpe.[8] Wyclif wanted to find a third way between two positions with respect to the reality of things referred to by universal concepts.[9] The first is Walter Burley's view (after 1324) that universals exist in the world independently of thought and that they are really distinct from the individuals in which they are present. In analysing the notions of identity and difference, Burley advances an influential explication of identity and difference which was later called the principle of the indiscernibility of identicals:

> If something is predicated of a thing of which another is not predicated, these are not the same, but different, and if something is predicated of a thing and it is not predicated of another, these are not the same. And contrarily, if anything which is truly predicated of one thing is truly predicated of another, they are the same.[10]

According to Burley, A and B are the same if exactly the same predicates can be truly said of them, and there is a real distinction between them if different predicates can be truly said of them. Distancing himself from Burley's real distinction between universals and their individuals, Wyclif

7 Ibid. 203-4.

8 See the entries for these by Alessandro Conti in *The Stanford Ecyclopedia of Philosophy* (http://plato.stanford.edu).

9 John Wyclif, *Tractatus de universalibus*, ed. Ivan J. Mueller (Oxford: Oxford University Press, 1985), 4.40-59, translated by Antony Kenny in John Wyclif, *On Universals* (Oxford: Oxford University Press, 1985), 27-8.

10 Walter Burley, *Expositio super artem Veterem Porphyrii et Aristotelis* (Venice, 1509), 44r.

also criticises another view which he dedicates to Thomas Aquinas and Giles of Rome. By Wyclif's lights, this approach involves extra-mental singular substantial things being instantiations of universal forms which are metaphysical constituents of things and real in this sense, without representing a different kind of existing thing. There is no mind-independent distinction between universals and their particular instantiations. Wyclif's own solution is that universals and their individuals are neither indiscernibly the same nor really different, since there is a formal distinction between a concrete individual and the corresponding universal form which occur in the world as numerically indistinct.[11]

In his introduction to Wyclif's *On universals*, Paul Vincent Spade explains Wyclif's theory of identity and distinction by taking numerical identity as a basic notion in terms of which one may define the notion of individual as follows: '*x* is an individual if and only if all the entities numerically identical with *x* together yield *one x*, not several'.[12] Every entity in Wyclif's universe is taken to be numerically identical with at least one individual; either the entity is itself an individual or else some kind of constituent of one or more individuals. Numerical identity is reflexive and symmetrical but not transitive. The species called human being is numerically identical with Socrates and Plato and vice versa, but while Socrates is numerically identical with the species and this with Plato, Socrates and Plato are not numerically identical, but numerically distinct, as are all individuals from each other. Individuals may inhere in other things in Wyclif's metaphysics, but there are also metaphysical subjects or substrata (supposita) which do not inhere in other entities. All entities which are not numerically identical are numerically distinct. Numerically identical entities may be formally distinct, in which case there is one suppositum, the distinction being between this and the entities it supports or between those supported entities. Formally distinct entities are numerically identical but discernible. In addition, the supposita, which are numerically identical with the entity which each of them alone supports, may really be distinct from each other, as with divine persons.[13]

In Wyclif's realism, predication is primarily a relation between things and true predications of terms are based on the more basic predications of things. This idea was applied by many of his followers, including Johannes Sharpe, who describes two conceptions of essential predication among the Oxford Realists. The first is that:

---

11 See also Paul Vincent Spade, 'The Problem of Universals and Wyclif's Alleged "Ultrarealism"', *Vivarium* 43 (2005), 120-1.

12 Paul Vincent Spade, 'Introduction' in John Wyclif, *On Universals* (note 9 above), xxvii.

13 Spade 1985, xxi-xxxi; Spade 2005, 122-3.

> All predications in things are essential, for in all such predications something the same or the same entity or essence is the subject and the predicate, in the same way as in all predications of signs which are rightly subordinated to these something the same or the same entity is signified by the subject and the predicate.[14]

Apart from the claim that essential predication is the basis of all predications, this was how essential predication was understood from Wyclif to Sharple and how Paul of Venice described identical predication.[15]

I referred above to a passage in which Paul of Venice discusses some logical problems about universals as the same as those associated with Trinitarian formulations – in fact he says that the realist view of universals with the distinctions just described was developed as an answer to theological queries.[16] One of these was the question of whether the doctrines of the Trinity and the Incarnation were compatible with the universal validity of logic. Two examples were discussed by most late medieval theologians and logicians, formulated by Roger Roseth (c. 1335) as follows:

> First it seems that no expository syllogism holds in divine matters, for this does not hold: This divine essence is the Father, this divine essence is the Son; therefore, the Son is the Father … further, in the first mood syllogisms are not valid, for this is not valid: Every divine essence is the Father, every Son is the divine essence; therefore every Son is the Father.[17]

These arguments seemingly denied two basic principles of syllogistic logic. The validity of the first figure moods was often explained by the *dici de omni et nullo* in medieval logic, which was regarded as a self-evident principle explaining the transitivity of inclusion and exclusion in first figure syllogisms.[18] The Trinitarian first mood syllogism seemingly violated the *dici de omni et nullo* principle; the first example questioned the validity of expository syllogism which was understood by many fourteenth-century authors as one with singular premises. Some authors regarded the new expository syllogism as another immediately evident syllogistic principle, while others wanted to reduce it to the *dici de omni et nullo*, treating singular premises as equivalent to corresponding universal premises. There were theologians to argue that Trinitarian paralogisms showed that a special *logica fidei* was needed to improve the traditional logic, which was not uni-

---

14 *Quaestio super universalia*, 90.

15 See also Alessandro Conti, 'Johannes Sharpe's Ontology and Semantics: Oxford Realism Revisited', *Vivarium* 43 (2005), 156-86.

16 *Quaestio de universalibus*, 200.

17 Roger Roseth, *Lectura super Sententias*, 3-5, ed. Olli Hallamaa (Helsinki: Luther-Agricola Society, 2005), 3.1 (67-8).

18 See, for example, Robert Kilwardby, *In libros Priorum Analyticorum expositio* (Venice, 1516, under the name of Giles of Rome, reprinted Frankfurt am Main: Minerva, 1968); 10rb, 11ra.

versally valid, but it was more usual to hold that these paralogisms and analogous arguments pertaining to universals and their instantiations were resolved by explicating the meaning of the terms and the logical structure of propositions in counter-examples, i.e., explaining the supposition of terms in theological statements and analysing the kinds of predication, such as essential or identical predication and formal predication.[19] This approach is also found in Wyclif's *On universals* and the works of his followers.[20] Next, I will explain some features of the historical background to these discussions.

2. Abelard on Identity and Predication

One of the central topics in Peter Abelard's logical works was the analysis of predication in grammar and logic. In the so-called *Logica Ingredientibus*, Abelard suggests that a proposition like 'A human being is pale' as logically equivalent to 'The same which is a human being is that which is pale'.[21] The copula indicates that the terms apply to the same subject – this is the kernel of the identity theory of predication. While the identity analysis was not the only idea of predication in Abelard, it was the logical cornerstone of his discussion of the question of how Trinitarian persons, being one and the same God, are distinct from each other.

Abelard calls the extensional numerical identity which is expressed by 'An A is B' an *idem quod* identity or *identitas essentiae*, as distinct from the intensional identity expressed by synonymous terms, which he calls *identitas proprietatis*.[22] In the first case, the terms name things which are numerically identical particular beings (*essentia*). Things can be same in this way without exhibiting all the same properties or attributes. Taking a waxen image as an example, Abelard says that the lump of wax and the waxen image are essentially the same, even though they have different properties,

19 See Michael H. Shank, *'Unless You Believe, You Shall Not Understand.' Logic, University, and Society in Late Medieval Vienna* (Princeton: Princeton University Press, 1988), 79-81, 89-96; Alfonso Maierù, 'Logic and Trinitarian Theology: *De modo predicandi ac Sylogizandi in Divinis*' in Norman Kretzmann (ed.), *Meaning and Inference in Medieval Philosophy*, Synthese Historical Library 32 (Dordrecht: Kluwer, 1988), 247-95; Simo Knuuttila, 'The Question of the Validity of Logic in Late Medieval Thought' in Russell Friedman and Lauge Nielsen (eds.), *The Medieval Heritage in Early Modern Metaphysics and Modal Theory 1400-1700*, The New Synthese Historical Library 53 (Dordrecht: Kluwer, 2003), 121-42.

20 *Tractatus de universalibus* I.210-34, 287-97.

21 Peter Abelard, *Logica 'Ingredientibus'*, ed. Bernhard Geyer, Beiträge zur Geschichte der Philosophie des Mittelalters 21.1-3 (Munich: Aschendorff, 1919-27), 60.8-13.

22 See *Theologia Christiana*, ed. Eligius M. Buytaert, Corpus Christianum Continuatio Mediaevalis 12 (Turnhout: Brepols, 1969), IV.36; IV.40-1; IV.46; IV.52-3; 56-7; IV.102.

as is clear from the fact that the wax, as distinct from the image, is not made from wax.[23]

It has been argued that Abelard's analysis is the same as what is called an Aristotelian answer to the problem of material constitution which arises when two objects appear to share all the same material parts and yet are related to their parts in different ways. In Abelard's example, the lump of wax and the statue share all the same parts though they are in some sense different. Abelard's position is described as follows. We count one *statue* in every region that is filled by matter arranged statuewise, and we count one *lump* in every region that is filled by matter arranged lumpwise, and we count one *object* in every region that is filled by matter arranged in either or both of these ways or any other object-constituting way.[24]

Abelard discussed the constitution problem with respect to material objects taking on some shapes or forms in hylomorphic compounds. Contrary to what one might expect, he did not try to modify this to cover immaterial objects. He regarded God as metaphysically simple, without the complexities associated with the possession of parts, forms or properties. Instead of speaking about the Trinity as numerically one entity with distinct constituents, Abelard preferred to speak about the persons of the Trinity as different on the basis of what was proper to them, each numerically same as the divine substance. This did not imply that one should speak about plural properties or forms in a metaphysical sense. While not using the constitution model as a metaphysical explanation of the Trinity, Abelard believed that the distinctions between the kinds of identity exemplified by this model offered a logically valid analysis of Trinitarian propositions.[25]

According to Abelard, many Trinitarian formulations were meant to express the identity of essence:

> It is merely the identity of essence and not that of property which is meant in these: the Father is God, the Son is God, the Holy Spirit is God ... for it was only shown that the thing referred to by the predicate term and the thing referred to by the subject term is the same; for example, the same which is God is that which is the Father. [26]

Abelard thought that this analysis largely solved the logical queries which might be associated with the Trinity, one of these being the question of

23 Ibid. III.140.

24 Jeffrey E. Brower, 'Trinity' in Jeffrey E. Brower and Kevin Guilfoy (eds.), *The Cambridge Companion to Abelard* (Cambridge: Cambridge University Press, 2004), 231-4.

25 The Abelardian type of constitution theory is applied to Trinitarian theology as an improvement of medieval and contemporary approaches in Jeffrey E. Brower and Michael C. Rea, 'Material Constitution and the Trinity', *Faith and Philosophy* 22 (2005), 57-76; see also Brower 2004, 235-43.

26 *Theologia Christiana* IV.52-3.

whether the propositions 'The Father is God' and 'God is the Son' syllogistically implied that 'The Father is the Son'. Abelard remarks that if the premises are read according to essential identity, the conclusion should be understood in the same way. This is not problematic, since all these propositions are true in this sense. If the conclusion is wrongly understood in the sense of intensional identity, it does not follow from the premises, which are true merely in *the* sense of essential identity.[27]

3. John Duns Scotus and Adam Wodeham

The Abelardian distinction between intensional and extensional identity was included in twelfth-century theology with some modifications. In Trinitarian discussions, the *idem quod* predication was associated with terms interpreted as *substantive* and regarded as referring to one and the same suppositum. This reading of Trinitarian propositions was often true, even when treating the predicate *adiective* and directly predicable of the subject was false, for example 'The essence generates'. Bonaventure called the former predication *per identitatem* and the latter predication *per inherentiam*.[28] Bonaventure's distinction was often mentioned by later authors and was also known to John Duns Scotus who introduced the related distinction between identical and formal predication (without ever defining these terms).

In dealing with the expository syllogism example, John Duns Scotus employs the often-used phrase 'things that are the same as one and the same thing are also the same as each other' and the distinction between two kinds of identity. According to him, when two divine persons are said to be the same as the essence, this can be understood in the sense of essential identity (*identitas essentialis*) – the medium is one essence which is communicated to two subjects, and consequently the correct conclusion is 'The Father is the same as the Son', namely, the single essence. The premises and the conclusion are true when understood in this way. The false conclusion, 'The Father is the Son', which is said to express formal identity (*identitas formalis*) or suppositional identity (*identitas suppositiva*), follows from the premises which also express such an identity. 'The essence is the Father' and its converse are true, but there is a formal distinction between the Father and the essence, although these are numerically the same. The same holds of 'The essence is the Son'. Even though the persons are nu-

27 *Theologia Christiana* IV.56.

28 *Commentaria in quatuor libros Sententiarum*, I.5.1.1, ad 2-3 in *Opera omnia* I, ed. Collegium S. Bonaventurae (Ad Claras Aquas: Collegium S. Bonaventurae, 1882).

merically identical with one God and one divine essence, the persons themselves are numerically different supposita. The essence is a communicable individual with three incommunicable supposita. Scotus thinks that the mistaken syllogism could be based on not realizing that the essence is a metaphysically simple individual and also communicable like created common natures, albeit different from these by existing in itself as an individual.[29]

According to Scotus, there are two types of predication in Trinitarian theology: identical predication, which is based on essential identity, and formal predication, which is based on formal identity.[30] This is not the same as Abelard's distinction mentioned above in which *idem quod* sameness or essential identity pertains to the sameness of the subject of which the terms are expressed, the ultimate subject being one and the same in all Trinitarian propositions of the *idem quod* type. Scotus seems to think that the simple unity of divine essence is the basis of identical predication as a common predicate rather than as a common subject.

Leibniz described the counting of similar things as follows: 'If B is A and C is A and B and C are not the same, it is said that there are two As. And if it is added that D is A, and none of these B, C, and D, is the same as one another, it is said that there are three As, and so on.'[31] According to Scotus, if B and C are divine persons which are numerically same as the divine essence, there are two numerically distinct things which are the essence, not two essences. (Leibniz also mentions this in the same place.) The denial of the transitivity of numerical identity was the specific logical problem of the Trinity. Adam Wodeham, a student of William Ockham and an influential fourteenth-century theologian, tried to explain this by reformulating the 'sufficiently universal' form of a syllogistic universal proposition as follows: 'Everything which is A is the same as that which is B. 'Every A is B' is ambiguous because it may be taken to mean that A as a single common nature is numerically the same as those which are B. This is not a sufficiently universal syllogistic proposition which fulfils the requirements of the *dici de omni et nullo*, as the phrase 'which is A' explains. 'Every essence is the Father' is true on the former reading but false on the latter, since the persons, while numerically the same as the essence, are numerically distinct from each other. Wodeham argues that singular

29 *Ordinatio* I, 2.2.1-4, 403, 411-17 in *Opera omnia*, ed. Carolus Balić et al., vol. 2 (Vatican City: Typis Polyglottis Vaticanis, 1950); *Lectura* I, 2.2.1-4, 275, 278-82 in *Opera omnia*, vol. 16 (Vatican City: Typis Polyglottis Vaticanis, 1960). See also Richard Cross, 'Duns Scotus on Divine Substance and the Trinity', *Medieval Philosophy and Theology* 11 (2003), 181-201.

30 *Ordinatio* I, 8.1.4, 217-8, *Opera omnia*, vol. 4 (Vatican City: Typis Polyglottis Vaticanis, 1956).

31 *Notationes generales*, in G.W. Leibniz, *Sämtliche Schriften und Briefe*, VI.4, ed. Heinrich Schepers et al. (Berlin: Akademie-Verlag, 1999), 131 (550).

propositions should also be universalised in this way, which solves most of logical problems – many sufficiently universal and transitivity supporting Trinitarian premises proves to be false. This is associated with the additional problem that among things of which A is correctly said there may be things which the term A directly supposits for as well as things which are numerically same as these, although A does not directly supposit for them; therefore, in a full explication of a Trinitarian affirmative proposition, 'which is A' should be understood as 'which is A or is the same as that which A' and similarly with the predicate term. In order to avoid the impressions that these are merely *ad hoc* theological considerations, Wodeham argues that similar ideas could be applied to the Platonic ontology of real common entities.[32]

Many early fourteenth-century logicians put forward an identity theory of predication which showed similarities to Abelard's view. William Ockham and John Buridan also mentioned that the explication 'which is A' of 'Every/some/one A' was required to avoid Trinitarian counter-examples to syllogisms.[33] Accepting the identity theory of predication, Wodeham argued that the above analysis was sufficient against all counter-examples. These widely-accepted ideas were formulated by Roger Roseth as follows:

> One principle is that not every proposition is sufficiently singular to be a premise in an expository syllogism, for in order to be sufficiently singular it should be convertible to and equivalent to a universal proposition in which the subject of such a singular proposition is circumlocutioned, for example, as this proposition 'This human being runs' is equivalent to 'Everything which is this human being runs' ... The second principle is that a universal affirmative proposition can be a premise of a syllogism which is regulated by the *dici de omni* only if its subject is equivalent to a circumlocuted subject; for example ... these are converted to each other 'Every human being runs' and 'Everything which is a human being runs'.[34]

This was the historical background of the discussion of various numerical identities in logic and theology until Leibniz.

---

32 *Super quattuor libros Sententiarum*, abbreviated by Henry Totting of Oyta (Paris, 1512), I, 33.3 (81ra-vb).

33 John Buridan *Tractatus de consequentiis*, ed. Hubert Hubien, Philosophes médiévaux, 16 (Louvain: Publications Universitaires; Paris: Vander-Oyez, 1976), III .1.4.1 (86); William Ockham, *Summa logicae*, ed. Philoteus Boehner, Gideon Gál, Stephen Brown, *Opera philosophica*, vol. 1 (St. Bonaventure, N.Y.: St. Bonaventure University, 1974), III-1.4 (370-1).

34 *Lectura super Sententias*, 3.1 (71).

4. Identity in Temporal and Modal Predication

Perhaps the most important achievement of late medieval logic was the new theory of modality which was based on the systematic division between modalities *de dicto* and *de re*, the division of *de re* modalities into those having actual subjects and those having possible subjects, and the interpretation of possibility with the help of the conception of simultaneous alternatives. One of the architects of the new modal theory was John Buridan.[35] Let us see how Buridan applies the identity theory of predication to singular tensed and modalized propositions:

> Furthermore, it is also clear that if we say 'A is B', then provided that the terms are not ampliated to the past or future, it follows that to say 'A is B' is equivalent to saying that A is the same as B – and thus also to say that A is not B is equivalent to saying that A is not the same as B. But if it is true that A is the same as B, then it is necessary that the terms 'A' and 'B' supposit for the same thing, in the sense that some A should be posited to be the same as some B. And the same goes for the past and the future. For there is no difference in saying 'Aristotle was someone disputing' and 'Aristotle was the same as someone disputing'; therefore, in this proposition the terms 'Aristotle' and 'someone disputing' supposit for the same … not because Aristotle and someone disputing are the same, but because they were the same, and the case is similar with the future and the possible.[36]

In past or future tense propositions, the terms stand for past or future objects of which it is maintained that they were or will be the same, or, if the subject term is not ampliated, then something actual was or will be the same as something. In *de re* possibility propositions, either both of the terms or the predicate term stand for possible beings and these are said to be possibly the same. 'The present pope was a child' is read 'This was the same as a child' rather than 'This is the same as a past child' and, correspondingly, if one speaks about the seated Socrates' simultaneous possibility of standing, this is analysed as 'This is possibly a standing one' rather than as 'This is the same as a possible standing one'. Of an actual being one might say that it was or will be or can be the same as something which is past or future or possible in comparison to actual states of affairs. This is a form of actualism in which the domains of the past, future and possibility are regarded as real even though they do not exist. The supposita of temporally ampliated terms include past and future subjects and those of modally ampliated terms also include merely possible subjects; all these are treated as demonstrable by demonstrative pronouns, although

35 Simo Knuuttila, 'Medieval Modal Theories and Modal Logic', in Dov M. Gabbay and John Woods (eds.), *Handbook of the History of Logic*, 2 (Amsterdam: Elsevier, 2008), 551-559.

36 John Buridan, *Sophismata* II, tenth conclusion, in John Buridan, *Summulae de Dialectica*, trans. Gyula Klima (New Haven: Yale University Press, 2001).

this does not mean concrete demonstrability because many of these do not exist. One can speak about the same subjects in temporally separated situations and in various counterfactual states of affairs – these are the basis of tensed, contingent, or necessary *de re* predications. If an A is necessarily B, what is A cannot be consistently imagined without B, and it is contingently B, if it can be imagined with and without B.[37]

Many historians of logic have attended to similarities between fourteenth-century modal logic and the modern possible worlds semantics. One of these is the idea just mentioned that one can consider the same individual in alternative states of affairs. When Leibniz assumed that things merely have counterparts in alternative worlds, this was not a medieval view, as can be illustrated by Gilbert of Poitiers's twelfth-century definition of Plato's Platonitas, which involves everything Plato is, was or will be as well as whatever he could be without ever being it. This perfect concept of the individual is compatible with fourteenth-century modal theory.[38]

The question of identity in opaque contexts was extensively discussed in late medieval epistemic logic. It was usually thought that knowledge statements *de dicto* did not imply knowledge statements *de re* or *vice versa.* Buridan says, however, that when a person knows that some A is B, then of something which is A he or she knows that it is B. The reason for denying this could be that Socrates does not know which A is B. Buridan would agree that in this sense the *de re* reading does not follow from the *de dicto* reading, but there is another kind of *de re* reading (or one which might be called so) which does follow from the *de dicto* reading. According to Buridan, statements of the type

$$K_s(Ex)(Fx)$$

imply that there are individuals with property F, although Socrates does not necessarily know which they are. In principle they are identifiable, however, and if we suppose that one of them is z, we can write:

$$K_s (Ex)(Fx) \rightarrow (Ex)((x = z) \& K_s (Fx)).$$

From the *de dicto* statement 'Socrates, who is sitting in a cellar, knows that a star is above' it does not follow the *de re* reading understood as 'There is a star which Socrates knows as the star which is above', but the following *de re* reading does follow from it: 'There is a star of which Socrates knows that it is above, although Socrates does not know which star it is.'[39]

37 Knuuttila 2008, 555-6.

38 Knuuttila 2008, 521-2.

39 Buridan, *Sophismata* IV, 14 in *Summulae de Dialectica*, 900-2; Knuuttila 2008, 562.

Primary sources

Adam Wodeham, *Super quattuor libros Sententiarum*, abbreviated by Henry Totting of Oyta, Paris, 1512.

Bonaventura, *Opera omnia* I, ed. Collegium S. Bonaventurae, Ad Claras Aquas: Collegium S. Bonaventurae, 1882.

Johannes Sharpe, *Quaestio super universalia*, ed. Alessandro Conti, Unione Accademica Nazionale, Corpus philosophorum medii aevi, testi e studi 9, Florence: Olschki, 1990.

John Buridan, *Summulae de Dialectica*, trans. Gyula Klima (New Haven: Yale University Press, 2001).

John Buridan, *Tractatus de consequentiis*, ed. Hubert Hubien, Philosophes médiévaux, 16, Louvain: Publications Universitaires; Paris: Vander-Oyez, 1976.

John Duns Scotus, *Opera omnia*, ed. Carolus Balić et al., Vatican City: Typis polyglottis Vaticanis, 1950 - .

John Wyclif, *Tractatus de universalibus*, ed. Ivan J. Mueller, Oxford: Oxford University Press, 1985; *On Universals*, translated by Anthony Kenny, Oxford: Oxford University Press, 1985.

Leibniz, G.W., *Sämtliche Schriften und Briefe*, vol. VI.4, ed. Heinrich Schepers et al., Berlin: Akademie-Verlag, 1999.

Leibniz, G.W., *Opuscules et fragments inédits de Leibniz*, ed. Louis Couturat, Paris: Felix Alcan, 1903.

Peter Abelard, *Logica 'Ingredientibus'*, ed. Bernhard Geyer, Beiträge zur Geschichte der Philosophie des Mittelalters 21.1-3, Munich: Aschendorff, 1919-27.

Peter Abelard, *Theologia Christiana*, ed. Eligius M. Buytaert, Corpus Christianum Continuatio Mediaevalis 12, Turnhout: Brepols, 1969.

Robert Kilwardby, *In libros Priorum Analyticorum expositio*, Venice, 1516, under the name of Giles of Rome, reprinted Frankfurt am Main: Minerva, 1968.

Roger Roseth, *Lectura super Sententias*, 3-5, ed. Olli Hallamaa, Helsinki: Luther-Agricola Society, 2005.

Walter Burley, *Expositio super artem Veterem Porphyrii et Aristotelis*, Venice, 1509.

William Ockham, *Summa logicae*, ed. Philoteus Boehner, Gideon Gál, Stephen Brown, Guillelmi de Ockham Opera philosophica, vol. 1, St. Bonaventure, N.Y.: St. Bonaventure University, 1974.

Secondary Sources

Antognazza, Maria Rosa (2001) 'The Defence of the Mysteries of the Trinity and the Incarnation: An Example of Leibniz's 'Other' Reason', *British Journal for the History of Philosophy* 9, 283-309.

Brower, Jeffrey E. (2004) 'Trinity', in Jeffrey E. Brower and Kevin Guilfoy (eds.), *The Cambridge Companion to Abelard*, Cambridge: Cambridge University Press, 223-257.

Brower, Jeffrey E. and Michael C. Rea (2005), 'Material Constitution and the Trinity', *Faith and Philosophy* 22, 57-76.

Conti, Alessandro (2005), 'Johannes Sharpe's Ontology and Semantics: Oxford Realism Revisited', *Vivarium* 43, 156-186.

Cross, Richard (2003), 'Duns Scotus on Divine Substance and the Trinity', *Medieval Philosophy and Theology* 11, 181-201.

Knuuttila, Simo (2003), 'The Question of the Validity of Logic in Late Medieval Thought', in Russell Friedman and Lauge Nielsen (eds.), *The Medieval Heritage in Early Modern Metaphysics and Modal Theory 1400-1700*, The New Synthese Historical Library 53, Dordrecht: Kluwer, 121-142.

Knuuttila, Simo (2008), 'Medieval Modal Theories and Modal Logic', in Dov M. Gabbay and John Woods (eds.), *Handbook of the History of Logic*, 2 (Amsterdam: Elsevier, 2008), 504-578.

Lenzen, Wolfgang (2004), *Calculus universalis. Studien zur Logik von G.W. Leibniz*, Paderborn: Mentis.

Maierù, Alfonso (2003), 'Logic and Trinitarian Theology: *De modo predicandi ac Sylogizandi in Divinis*', in Norman Kretzmann (ed.), *Meaning and Inference in Medieval Philosophy*, Synthese Historical Library 32 (Dordrecht: Kluwer, 1988), 247-295.

Shank, Michael H. (1988) *'Unless You Believe, You Shall Not Understand.' Logic, University, and Society in Late Medieval Vienna*, Princeton: Princeton University Press.

Spade, Paul Vincent (1985), 'Introduction', in John Wyclif, *On universals*, vii-l.

Spade, Paul Vincent (2005), 'The Problem of Universals and Wyclif's Alleged "Ultrarealism"', *Vivarium* 43, 111-123.

TIME AND EXISTENCE

LUDGER HONNEFELDER

The different meanings of the word “to be” have interested philosophers at least since Aristotle. In my paper, I would like to draw attention to the following question: in what way “to be”, when predicating actual existence of an individual, is connected with time and how it refers to the individual and its real change. I want to do this by presenting Thomas Aquinas’ interpretation of the Aristotelian analysis of the different meanings of the word ”to be” referring to Peter Geach, who discussed this doctrine in relation to the interpretation of “existence” by Russell and Quine.[1] My aim is to ask whether this medieval doctrine may contribute to the clarification of the relation between unity and time with regard to the question of endurantism and perdurantism.

I.

In Perihermeneias ch. 3 Aristotle states:

> “A verb is that which in addition to its proper meaning carries with it the notion of time ... That it carries with it the notion of time means that ‘health’ is a noun, but ‘is healthy’ is a verb; for besides its proper meaning it indicates the present existence of the state in question.”[2]

The indication of time, Aristotle explains, is not added to the verb as a word uttered by itself, but only in combination with other expressions, “for neither ‘to be’ or ‘being’ is significant of any fact, unless something is added”.[3] The fact that the verb in the sentence as an inflected form “carries with it the notion of time” by its tense, is trivial. But to say that the word “is”, used in the predicate-position, has to be understood as a verb of the same kind as “runs” and that it carries with it something else when used in this way, not only is not trivial, but – as early analytical philosophy asserts – false.

---

1 Conf. also my former interpretation in: L. Honnefelder, ‘Zeit und Existenz’, 333-362.

2 Aristotle, *De Interpretatione*, c. III, 16 b 6-9.

3 Ibid., c. III, 16 b 23f.

According to Russell's well-known thesis "a exists"[4] is a phrase which is grammatically correct, but logically misleading. The absurdity is removed, Russell suggests, if we no longer understand 'a exists' as a subject-predicate connection like 'a runs', i.e. not at all as a *predicative* proposition, but as a statement asserting that the designation for which 'a' stands is not empty.

But, as we know, Russell's objections are for their part based on premises to which alternatives can be conceived. Nevertheless, what makes Russell's criticism right, is the peculiar status of the proposition 'a exists'; by its status it is distinguished from all other predicative propositions.

This peculiar status has been identified much earlier. The eleventh-century-philosopher Ibn Sina, or, in the Latin version of his name: Avicenna, and the authors who follow him emphasize the quasi-accidental character of 'est' in the existential statements. [5] Aquinas, in connection with the Aristotelian analysis of the different meanings of "to be", acknowledges the distinction between 'is' in the *quantificational logical* and in the *predicative* sense.[6] John Duns Scotus says about the 'est' of the existential statement that it is a predicate outside the signification of the essential properties, the "coordinatio praedicamentalis", insofar as it predicates only the "last act" of being outside its causes, the "esse extra causas".[7] And it is this doctrine which is taken up by Kant's thesis according to which 'being' in the sense of existence is a predicate sui generis, not a "real" predicate, i.e. not one which determines the quiddity (or "whatness") of a thing, but a predicate that sets "merely the position of a thing, or of certain determinations, 'as existing in themselves'".[8] In a similar way, Frege says that existence is not a criterion, but a property of a concept and has therefore not to be considered as a first-level, but as a second-level concept.[9]

The benefit of this solution with regard to the interpretation of "is" in the sense of "exists" is obvious. "Existence" not only loses its ambiguity and inconsistency, but it can also be used now in a very unrestricted sense. Since there is nothing – as far as we can say anything about that at all –

4 Conf. B. Russell, 'On Denoting', 41-56; B. Russell, 'The Philosophy of Logical Atomism', 177-281.

5 Conf. Avicenna, *De prima philosophia* VIII, 4-7, 397-433.

6 Conf. Thomas Aquinas, *Summa Theologiae* I 3, 4, 2.

7 Conf. John Duns Scotus, *Ordinatio* II d.3 p.1 q.3 nn. 53-65, 419-421. Conf. L. Honnefelder, *Scientia transcendens*, 140-158.

8 I. Kant, *Critique of Pure Reason* A 599 ff./B 626 ff. Conf. L. Honnefelder, *Scientia transcendens*, 469-472.

9 Conf. G. Frege, 'Dialog mit Pünjer über Existenz', 69.

which cannot appear as a logical subject, the predicate "there is", related to the logical subject, can be attributed to absolutely everything, irrespective of the species or category it belongs to. From a philosophical perspective, this is anything but trivial. For it allows us to say, in a semantically meaningful way, that beings that transcend our experience of space and time, such as God, exist. According to Aquinas, the only way to say of God, whose being and nature we are unable to know, that He is, lies in asserting – by inferring from effects to their causes – that the proposition "Deus est" is true.[10] The proof of the existence of God lies in demonstrating that the predicate "God", or whatever we use instead, is not empty and that the phrase built by using it is true. "Something existing is God", Kant says in a similar way, "that means that those predicates that we designate collectively by the expression 'God' belong to an existing thing."[11]

But being generalised this way, the predicate "exists" has not only lost its inner reference to the modes of being or existing, but also its inner connection with time, which had been claimed by Aristotle. If a temporal reference occurs in a existentially quantified proposition, it refers to the spatio-temporal location to which "there is" applies, but not to "there is" itself. Propositions which can be formalised according to canonical logic of predicates are always true (if they are true) or always false (if they are false), regardless of the point in time they relate to.

II.

But precisely the existential statements which Russell thought he could understand as tautological and which Quine thought he could avoid entirely by transforming them into a description, namely the singular existential statements[12], resist all attempts of becoming reduced to the existential quantifier and show, on closer inspection, that temporal reference is a factor that can by no means be eliminated. This irreducibility becomes evident in singular propositions, in which a proper name takes the place of the subject and in which the temporal reference of the predicate 'exists' is unmistakable: 'Until yesterday Peter existed, today he no longer exists.' Any transformation of the proper name into a description, even if it were 'to be Peter', may refer to the subject as an instantiation of a description, but not as a specific individual entity as a whole. If a proper name is concerned, it is impossible, according to Frege, to transform it into a predi-

10 Conf. footnote 6.

11 I. Kant, *The only possible Argument in Support of a Demonstration of the Existence of God* A 11.

12 Conf. W.V.O. Quine, 'Existence and Quantification'.

cate[13], as required by the quantificational logical interpretation of singular existential propositions. "When Mr. N.N. dies", Wittgenstein says, "we say that the bearer of the name dies, not that the reference dies."[14] The reference of the individual's name may not simply be identified with its bearer.

The quantificational rephrasing not only mistakes the way the proper name is used in the subject-position of our example; it fails utterly because of the predicative way in which 'exists' is used here. As Geach has demonstrated[15], to say that an individual came to exist, still exists or no longer exists, has a different meaning than if we say that the predication 'a exists' is false at a time t_1, true at a time t_2 and false again at a time t_3. Only if we understand 'exists' in the sense of individual existence, it is possible to say that something begins or ends. Apparently, 'exists' does have a predicative meaning of some kind when it is referring to persisting individuals. Kenny thus speaks of "individual existence"[16], which is to be distinguished from what he calls "specific existence" that can be expressed in terms of quantificational logic, and Tugendhat speaks of "temporal existence"[17], in contrast to the tenseless "exists" of the existential quantifier. Aquinas also bases himself upon this meaning when he interprets the "est" of singular existential statements like "Socrates est" as "actu est" and distinguishes it from the already mentioned "est", which has to be understood in terms of quantificational logic and which indicates the "veritas propositionis" in statements like "Deus est".[18]

III.

But how is "exists" in the case of singular existential propositions, which is used in a predicative sense and has a temporal reference, to be understood if it eludes integration into quantificational logical standard interpretation, and how can the aporias be avoided that are inextricably linked to such a use, according to Russell and Quine?

---

13 Conf. G. Frege, 'Über Begriff und Gegenstand', 75.

14 L. Wittgenstein, *Philosophical Investigations*, § 40.

15 Conf. P. T. Geach, 'Form and Existence', 42-64, especially 53-64; P. T. Geach, 'What Actually Exists', 65-74; P. T. Geach: 'Aquinas', 88-100.

16 A. Kenny, *Aquinas*, 50.

17 E. Tugendhat u. U. Wolf, *Logisch-semantische Propädeutik*, 197 ff.; E. Tugendhat, *Vorlesungen zur Einführung in die sprachanalytische Philosophie*, 468 ff.

18 Conf. footnote 6; conf. H. Weidemann, "'Socrates est"/"There is no such thing as Pegasus"', 42-59.

For Geach[19] the decisive starting point is the distinction that Aquinas makes between the "est" of the *veritas propositionis* and the "est" in the sense of *actu esse* and that later leads Frege to differentiate between "actuality" as an attribute of individual objects and "existence" in the sense of "there is ..." as an attribute of concepts.

According to Geach, sentences like "Peter existed, now he no longer exists", as well as sentences of the kind "Sokrates walked and now he runs" cannot be interpreted by means of the existential quantifier, because this implies – as Geach puts it, in allusion to McTaggert and Russell – only the "Cambridge change". Interpreted by means of the existential quantifier and expressed with regard to Socrates, the statement would say that "Socrates runs is true at t_1 and false at t_2". What is thus lost is the continuity within change, i.e. that which makes the change of Socrates a "real change". This real change is only expressible if we understand "existence" or "change" as a property of the respective individual object.

But, with regard to "existence", is not precisely this the source of the problems that have been mentioned? According to Geach the paradoxicality which is, following Russell and Quine, related to the interpretation of "existence" as a predicate or property, only arises, if we ignore the specific characteristic of the property of existence: it does not designate a particular property like 'wisdom', nor a particular activity like 'running', but simply refers to the *actuality* of those properties and activities that make a thing what it is; in doing so, it does not refer to the occurence here and now, which is expressible in terms of quantificational logic, but to the ongoing process of performing or realizing the properties that make a thing what it is, or, to say it the Aristotelian way, the actuality of being in one of the categories, or, as Aquinas puts it, the "esse" which has to be attributed to a thing in virtue of its individualised "forma". The temporal individual existence refers to the actuality or reality of a thing, insofar as this thing "acts, undergoes change or both"[20], or in short "actuality" in the sense of "actively or passively partaking in causal developments".[21] Thus, we must necessarily distinguish between that which is merely possible in thought and that which is actual: the ability to undergo change can not be attributed to something conceivable, but only to something actual.

To understand existence in this sense as a 'property' of the individual thing, thus means to express neither something tautological nor something nonsensical. And it does by no means involve the second contradiction claimed by Russell and Quine, according to which, in the case of the

19 Conf. footnote 15.

20 P. T. Geach, 'What Actually Exists', 65.

21 P. T. Geach, Art. Existenz, 206.

negation in the subject term, the existence of the individual is affirmatively asserted, but is then negated in the predicate term. Russell's contradiction only arises if the reference of the individual's name is identified with its bearer, as has been mentioned before. Thus, the sentence "Peter no longer exists" – if it is not supposed to be nonsensical – must not be read in a quantificational logical way, in the sense that among the things that are present here and now there is none which is Peter. It may be read predicatively, in the sense that the bearer of the name has ceased to be actual as the one to whom the name refers to. In this interpretation, existence appears as a 'property', but as a property sui generis, i.e. a property that is nothing but the actuality of that which makes the thing be what it is, in the same way as Aristotle states "vivere viventibus est esse" the being of living beings is nothing but their actual life. For something exists in the sense of individual existence, if it actualises the properties that correspond to its specific nature, or, as Aquinas says: "quodlibet esse est secundum formam aliquam".[22] The "actus essendi" – as Aquinas calls the act of individual existence – is indeed an act, but not an act of proper content, but an act in virtue of the form which is characteristic of the species.

This act only becomes manifest to us in the process of change, a change in fact that is understood as a property of the individual thing, as the example given above has shown. But if the change has to be understood as "real", the time implied can only be reconstructed as part of the predicate, speaking in terms of logic of language. Time appears as a constituent of existence of the individual things in the world we are experiencing.

IV.

But what does this mean for the connection of existence and time? If according to Geach, the ability to undergo real change is a criterion of individual existence[23], of course, this cannot mean that only that exists which can call a potency of this kind its own. For numbers, classes, etc. do not possess the potency to undergo real change. On the other hand, they do not commit us to attribute the aforementioned individual existence to them, but only the 'there is' of the existential quantifier.

But what about existence in case of individually existing things? Do we have to equate existence with the ability to undergo real change? If this

22 Thomas Aquinas, *Summa Theologiae* I 5, 5, 3.

23 Conf. footnote 20.

question has to be affirmed, this would mean, provided that real change is impossible without time and that time is impossible without the intrinsic order of earlier and later, that the extension over time phases is inextricably linked to the meaning of individual existence. In an exaggerated sense, individual existence and time would be synonymous.

Geach himself does not feel obliged to come to this conclusion. For he counts as 'acting', which he cites as a criterion of individual existence, both the "initiation of changes in things", and the "inner activities of mind like thinking and planning".[24] According to Geach, such activities of mind may by all means be understood as acts that initiate real change in things, but do not themselves represent real change. This implies that a concept of actuality can be formed which refers to a mode of process without real inner change, and that means without temporal extension. If there is someone whose existence solely consists in actuality of such a kind, we could say that it is "eternal".

But how can we refer to existence in the sense of actuality, without associating this with any notion of duration, and that certainly means, of successive states in time? If we do not pretend to have a privileged knowledge of eternity or if we do not want to dismiss the concept as nonsensical, because inexplicable, the only way to explain the concept of tenseless actuality is to relate it to actuality that takes place within time, i.e. to the real changes. Aquinas's candidate for such a reference is "simultaneity", understood as co-existence with the different temporal states of things that undergo real change. By means of this simultaneity, we may ascribe a "duration" to the tenseless thing, without being obliged to attribute real change and thus temporality to it.

Admittedly, this calls for a concept of simultaneity which is not, in analogy to the "proximity in space", that of a transitive relation. An understanding of eternity as "simultaneity to t_1, t_2, ...t_n" is indeed bound to expose itself to the incoherence which has been stated by Kenny. For if, in the sense of such a relation, A is simultaneous with t_1 and A is simultaneous with t_2, the same must apply to t_1 and t_2, which by definition is *not* the case. Simultaneity – Geach proposes - has therefore to be understood as being quasi non-relational, topic-neutral, in the same way the conjunction "while" links two propositions.[25]

If the concept of an actuality that is not real change can be presumed, a new finding results with regard to the relation of time and existence: although we may only introduce individual existence in the sense of actuality as a predicate by means of real change, 'existence' does not necessar-

24 Conf. Ibid.

25 Conf. P. T. Geach, 'Some Problems About Time', 311 f.

ily imply "time", with the exception of spatio-temporal things. This would however correspond to the aforementioned approach according to which "exists", in the case of individual existence, states actuality in virtue of the individual's form. Taking up the Aristotelian "vivere viventibus est esse", we could thus say of the eternal being that "existence" means an actuality which covers the totality of all conceivable states of "life"[26] outlined by "simultaneity" representing the "interminabilis vitae totae simul et perfecta possessio".[27]

V.

Something else results from Aquinas' analysis: if the quantificational logical "there is" represents states in a more general and wide sense, statements of individual temporal existence seems to be in some respects more fundamental. If Strawson's thesis of reference by identification is accurate[28], then Quine's reduction to general terms becomes impossible. Singular terms however imply the existence of that which is named, and in the case of singular terms this means individual existence of temporally existing things.[29]

Precisely this view is expressed by Aquinas, when he states a relation of dependency between the "is", in the sense of "being true of something", and the "is" of actual existence. The comment on the passage from Perihermeneias chapter III,[30] that has been quoted in the beginning, states that primary significance is being accorded to the "est" by which the "in actu esse" is attributed to a subject, in the sense of "actuality in virtue of the specific form" (actualitas omnis formae). However, if actuality is attributable to every being in virtue of the form, but is not simply identical to the form, this "est", in its primary significance, designates a "composi-

26 Conf. A. Kenny, *Aquinas*, 59.

27 Boethius, *De consolatione philosophiae* V 6.

28 P. Strawson, *Individuals*, 15-38.

29 Conf. E. Tugendhat, *Vorlesungen zur Einführung in die sprachanalytische Philosophie*, 464-471.

30 Conf. Thomas Aquinas, *In Peri Hermeneias* I, 5, n. 73: "Ideo autem dicit quod hoc verbum EST consignificat compositionem, quia non eam principaliter significat, sed ex consequenti; significat enim primo illud quod cadit in intellectu per modum actualitatis absolute: nam EST, simpliciter dictum, significat in actu esse; et ideo significat per modum verbi. Quia vero actualitas, quam principaliter significat hoc verbum EST, est communiter actualitas omnis formae, vel actus substantialis vel accidentalis, inde est quod cum volumus significare quamcumque formam vel actum actualiter inesse alicui subiecto, significamus illud per hoc verbum EST, vel simpliciter vel secundum quid: simpliciter quidem secundum praesens tempus: secundum quid autem secundum alia tempora. Et ideo ex consequenti hoc verbum EST significat compositionem."

tion" (compositio), which consists of essentia as the individualised form and actuality as the "actus essendi", as it comes across in being "ens" in the sense of that "which (actually) is" (quod est). Since "est", in this primary sense, designates actuality in all its forms, it may "by inference" (ex consequenti) designate the actual "in-esse" of a form in its subject by "connecting subject and predicate as a copula"; in virtue of the primary significance, this directly (simpliciter) applies to the present tense and only indirectly to the other tenses.

This idea could be extended further: as regards a being which is, like the things of our experience of the world, characterised by the fact that essence and being represent a unity in it, but do not coincide – unlike in the case of God, as Aquinas presumes – the "actu esse" can only be predicated of it with regard to the form attributed to it, i.e. in the mode of composition. The linguistic form of the compositio results from a unity in the difference of being and essence. The same applies to time: for if a being, the essence of which is not at the same time the reason for its existence, is that being which can come to exist and cease to exist, i.e. a temporal being, its existence can only be expressed as "secundum praesens tempus". Temporality of existence may however only be expressed in the temporal reference of a proposition, if the judging subject, as Aquinas demonstrates in the judgement analysis[31], is characterised by a reflexive relation towards its own existence and its temporality.

VI.

What is the relevance of Aquinas' analysis regarding to the subject of unity and time?

1. The distinction between a quantificational and a predicative interpretation of existence is not new. It can already be found in the medieval interpretation of the Aristotelian analysis of "to be". This interpretation refers to the natural language use; nonetheless, the arguments mentioned deserve attention irrespective of this.

2. As can be exemplified in Aquinas, this analysis comes to the conclusion that there are propositions in which existence of individuals that are designated by proper names is predicated in a sense that cannot be expressed adequately by the existential quantifier "there is". The transformation into "there is"- sentences fails to express the beginning or the end of

31 Conf. L. Oeing-Hanhoff, Art. Abstraktion, Sp. 47-59.

existence. For it merely allows us – as Geach says – to understand change as 'Cambridge change', not as a change that has to be attributed to the individual itself, i.e. as a "real change" of a persisting individual.

3. In such sentences, the predicate "exists" does not predicate an additional, specific property of the individual, but – according to Aquinas – the actuality of those properties that make the respective individual what it is and without which the individual not only would lack one of its properties, but would not exist at all. Aristotle says already that the actual existence of a living being consists in nothing else than the actuality of the property which characterises a living being as a living being as a whole, i.e. to live: vivere viventibus est esse.

4. That which makes a living being of a certain species a living being of this species is, according to Aquinas, the "form" which can be attributed to all living beings of this species, i.e. the complex of properties and activities that make a thing what it is. Singular temporal existence thus means that the form (type), is actualised and instantiated by an individual (token). The individual, which actualises its specific form, is presumed to possess the real ability to actualise the properties and activities belonging to its species, including the succession of the steps of realisation.

5. This actualisation of properties may also take place in the form of mental acts which are not bound to space and time. Tenseless actuality of existence can be understood as simultaneity with all points in time. It is eternal in the sense that it can be thought of as being simultaneous (simul) with all individuals that undergo real change. This way according to Aquinas the concept of "interminabilis vitae totae simul et perfect possessio" can be formed.

6. There are entities (such as concepts, classes etc.) that do not have a singular temporal or an atemporal existence in the sense of actuality, but of which "there is" statements can be made. "There is-sentences" that refer to universals or laws require the singular existential statements in order to be verifiable and thus seem to presuppose an endurantism of individuals.

7. According to the presented analysis it seems that we have to make a distinction between individuals that actualise their nature themselves and other entities the existence of which is only predicated in the sense of "there is".

8. Regarding individuals which actually exist it could be said that in this case time is identical with real change, i.e. with the "actu esse". The reason is – according to Aquinas – that these individual beings have their "actu esse" only in the mode of beings in which essence and the act of being do not coincide and which therefore begin or cease to exist. This temporality as a specific mode of being can only be recognized by subjects who have a reflexive relation towards their own existence.

References

Aristotle, *De Interpretatione.*

Avicenna, *De prima philosophia*, ed. Van Riet II.

Boethius, *De consolatione philosophiae.*

Frege, G. (1969a) 'Dialog mit Pünjer über Existenz' in: G. Frege, *Nachgelassene Schriften*, ed. H. Hermes u.a., Hamburg.

Frege, G. (³1969b) 'Über Begriff und Gegenstand' in: G. Frege, *Funktion, Begriff, Bedeutung. Fünf logische Studien*, Göttingen.

Geach, P. T. (1963) 'Aquinas' in: G. E. M. Anscombe and P. T. Geach, *Three Philosophers*, Oxford 1963, 88-100.

Geach, P. T. (1969a) 'Form and Existence' in: P. T. Geach, *God and the Soul*, London, 42-64.

Geach, P. T. (1969b) 'What Actually Exists' in: P. T. Geach, *God and the Soul*, London, 65-74.

Geach, P. T. (1980a) Art. Existenz, in: *Handbuch wissenschaftstheoretischer Begriffe*, Bd. 1, Göttingen.

Geach, P. T. (1980b) 'Some Problems About Time' in: P. T. Geach, *Logic Matters*, Berkeley/Los Angeles, 302-318.

Honnefelder, L. (1990) *Scientia transcendens. Die formale Bestimmung der Seiendheit und Realität in der Metaphysik des Mittelalters und der Neuzeit (Duns Scotus — Suárez — Wolff — Kant — Peirce)*, Hamburg.

Honnefelder, L. (1993) 'Zeit und Existenz' in: H. M. Baumgartner (ed.), *Das Rätsel der Zeit*, Freiburg-München. 333-362.

John Duns Scotus, *Ordinatio* II d.3 p.1 q.3, ed. Vat. VII.

Kant, I., *Der einzig mögliche Beweisgrund zu einer Demonstration des Daseins Gottes / The only possible Argument in Support of a Demonstration of the Existence of God.*

Kant, I., *Kritik der reinen Vernunft / Critique of Pure Reason.*

Kenny, A. (1980) *Aquinas*, Oxford.

Oeing-Hanhoff, L. (1971) Art. Abstraktion, in: J. Ritter u. K. Gründer (eds.), *Historisches Wörterbuch der Philosophie,* Bd. 1, Basel/Stuttgart, Sp. 47-59.

Quine, W. V. O. (²1971) 'Existence and Quantification' in: W.V.O. Quine, *Ontological Relativity and Other Essays*, New York/London.
Russell, B. (1956a) 'On Denoting' in: B. Russel, *Logic and Knowledge. Essays 1901-1950*, R. Ch. Marsh (ed.), London, 41-56.
Russell, B. (1956b) 'The Philosophy of Logical Atomism' in: B. Russel, *Logic and Knowledge, Essays 1901-1950*, R. Ch. Marsh (ed.), London, 177-281.
Strawson, P. (1959) *Individuals: An Essay in Descriptive Metaphysics*, London/New York.
Thomas Aquinas, *Summa Theologiae*.
Thomas Aquinas, *In Peri Hermeneias*.
Tugendhat, E. u. Wolf, U. (1983) *Logisch-semantische Propädeutik*, Stuttgart.
Tugendhat, E. (1976) *Vorlesungen zur Einführung in die sprachanalytische Philosophie*, Frankfurt a.M..
Weidemann, H. (1979) '"Socrates est"/"There is no such thing as Pegasus" Zur Logik singulärer Existenzaussagen nach Thomas von Aquin und W.V.O. Quine', *Philosophisches Jahrbuch* 86, 42-59.
Wittgenstein, L (1953) *Philosophical Investigations*, Oxford: Blackwell.

AGAINST CONVENTIONALISM: POWERS AND FUNCTIONS

EDMUND RUNGGALDIER

Abstract

The identity- and persistence-conditions of an individual depend on its individual sortal or essence. I argue against the conventionalistic account of this thesis. There is a mutual dependence relation between the continuity of the sortal and that of the basic capacities, powers and functions of an individual. Capacities, powers and functions are real and should not be reduced to ways of speaking with a mere heuristic role.

Introduction

In the account of persistence there is an ongoing discussion between two different positions due to different interpretations of the 4-dimensional space-time-system. The two positions depend on different accounts of the nature of time.

According to the first position – called '*perdurantism*' – every real entity is spread out in time, i.e. composed of temporal stages/parts. Everything is like a process or an event and, thus, there cannot be diachronic identity. Most *perdurantists* reduce it to a kind of continuity relation among adjacent temporal stages. The continuity relation is weaker and is neither reflexive nor transitive. It allows for differences of degree.

According to the second position – called '*endurantism*' – there are also *endurers* besides events. Things – human persons included – go with time, and are *present*, i.e., in the 'now'. They are fully present at each moment of their existence and therefore ontologically different from events.

Perdurantism conforms to conventionalistic empiricism. In this tradition it is commonplace to assume that individuals – living beings and human persons included – have to be *constituted/constructed* from 'the given', i.e., the data of experience. The *constitution* of individuals is *conventional*, its constraints being only pragmatic. Consistent conventionalism assumes this for the temporal dimension too. There is no 'fact of the matter' about diachronic identity.

In this paper I shall mention some counterintuitive consequences of conventionalism, affecting a mere perdurantistic account of persistence as well. I shall assume that it is the *individual sortal* or *sortal token* to guarantee the identity through time viz. persistence of a given individual. The thesis has to be unfolded, however: What does the sortal consist of und how can it be specified?

Sortals and sortal expressions have been accounted for conventionalistically. But statements on genuine sortals viz. natural kinds convey empirical knowledge on the identity- and continuity conditions of the individuals falling under them.

The continuity conditions of living individuals depend on their capacities, powers and functions. This last thesis presupposes an ontology that includes not only endurers, but properties as well. The sortal continuity is given by the continuity of certain capacities, powers and functions. The continuity of the basic capacities, powers and functions of an individual, be it a living being or a machine, helps to identify its sortal und to re-identify it as the same at various instances of its existence.

Counterintuitive consequences of conventionalism

Due not only to empiricism but also to Kantianism, many philosophers stick to the conventionalist theses that diachronic identity is not 'given' but has to be constructed or constituted. It seems to be a common opinion – at least in the German-speaking world – that neither experience nor empirical knowledge force us to accept transtemporal identity as real.

The various versions of traditional conventionalism converge in the view that all individuals we speak of in every-day life are *conventional posits*. The individuals as units in the spatial and temporal dimension are results of linguistic and cultural conventions. This, however, does not imply that there are no constraints in forming temporal units; these constraints are, however, merely pragmatic – lacking any cognitive or ontological basis.

Carnap was the first to develop in detail a *constitutional system* sketching how our every-day objects are derived, viz. constructed, from a basis referring only to sense experience. This basis does not convey any knowledge on alleged temporal or tensed facts, since the phenomenal 'given' does not contain any information on persistence relations. Carnap, however, cannot exclude the temporal dimension of reality altogether: it has an impact on memory and memory allows us to distinguish between similar and dissimilar experiences. He therefore chooses the *recollection of similarity* (Ähnlichkeitserinnerung) between fundamental or elementary experiences as a *basic relation* of his constructional system. (Carnap 1928, §61)

By forming similarity classes of elementary experiences based on the recollection of similarity he derives or constructs via complicated steps *qualities*. He then sketches how to construct visual objects, i.e., the spatial and temporal units of our every-day life. He conceives objects as *bundles of qualities* remaining in a stable relation to each other throughout the duration of their existence. 'If, in a bundle of world lines which have been constructed … the neighbourhood relations remain at least approximately the same during a protracted stretch (of time), then the class of the corresponding world-points is called a visual thing.' (Carnap 1928, § 128)

Due to the mentioned conventionalistic assumptions, the qualities have to be assigned to the world points viz. lines of the 4-dimensional system solely on practical-pragmatic constraints. Quine hints at the problem of this decisive step: '…qualities should be assigned to point-instances in such a way as to achieve the laziest world compatible with our experience'. (Quine 1963, 40) Quine sees the weak spot of Carnap's constructional procedure precisely in the connective 'quality q is at x, y, z, t' i.e. at a given point in the 4 dimensional space-time system. 'The connective 'is at' remains an added, undefined connective…' (Quine 1963, 40) Let us thus ask: are the factual constraints, excluding alternative objects, really only pragmatic? Aren't they limits due to our experience and knowledge of the world?

The conventionalistic principles of simplicity and utility alone are not sufficient to exclude the construction of the most fanciful objects that have nothing in common with those of 'our world'. (Runggaldier 1984, 140)

In spite of his critique of Carnap's conventionalistic connective that a quality is at a particular space-time-point, Quine too is conventionalistically minded in the account of the individuation of objects viz. of the spatial and temporal units of every-day life. The difference is due to his realistic conception of the distribution of stuff in the 4-dimensional space-time. It is a matter of fact and thus 'given' that the basic elements of reality are spread out in space-time as they are. But what regions of the space-time-matter we choose as individuals is conventional. Thus, our every-day objects are, for Quine, conventional *posits*. We can form scattered objects as well, i.e. objects interrupted in time. An object 'comprises simply the content, however heterogeneous, of some portion of space-time, however disconnected and gerrymandered.' (Quine 1960, § 36)

Each persisting concrete object should be thought of as a '*space-time worm*', a thing spread out in all four dimensions. 'Physical objects, conceived thus four-dimensionally in space time, are not to be distinguished from events or, in the concrete sense of the term, processes.' (Quine 1960, § 36) Thus, there cannot be diachronic identity: Nothing is destroyed when stuff is rearranged, neither by putting things together, nor by destroying them. The last consequence of a conventionalist approach to the problem of persistence concerns our own personal identity: nothing cognitively relevant forces us to think of the identity of our own self in the way we do. (Quine 1960, § 36)

However, a strict conventionalistic account of persistence does cause perplexities. One doubts whether really nothing factual constrains us in our statements on the persistence of objects. I will mention a thought experiment by E. Hirsch, in order to unfold some counterintuitive consequences of conventionalism.

Based on the conventionalistic assumptions it should be possible to form *alternative person*s combining 'temporal stages' in a different way. E. Hirsch has, thus, tried to see how such a possibility looks like by putting forward an alternative language differing from English solely in the formation rules of persons through time.

The idea behind the alternative language which Hirsch calls 'Contacti' is this: When two persons come into physical contact, each takes over the physical and mental characteristics of the other, and then, when they cease to be in contact, they regain their own characteristics. Thus, they have to use personal pronouns differently, even though the rue for the use of 'I' is the same, i.e., that a person must use it to refer to herself. (Hirsch 1982, 287ff)

In Contacti the term 'person' will not refer to persons in our sense, but rather to persons who swap their 'temporal stages' during their contact. Let us suppose that A and B come into physical contact with each other, then A becomes B and B becomes A. The histories therefore of persons in our every-day assumptions do not coincide with the histories of the persons in Contacti forming sums of different temporal stages. The history of the alternative A will not contain the stages of our A when it is in contact with B, but instead of these, the stages of B. The same applies analogously to the history of the alternative B.

If A is a man and B a woman, then during their embrace, the man would have to say in Contacti 'before the embrace I was a woman' and the woman could say in turn 'I was a man'. When A felt pain before the embrace, during the embrace he would have to say 'the women whom I am embracing was feeling pain and I can well remember what that feeling was like.' And she would have to say 'I was feeling pain a few moments ago, or so I'm told' (Hirsch 1982, 290).

In Contacti, the causal relations responsible for our memory and the connections between successive temporal stages of the life of persons have to be described in a radically different way. The peculiar concept of personal persistence presupposed by Contacti requires that the causal chains constitutive of memory can traverse the boundaries between one person and another. All this sounds counterintuitive.

Hirsch himself takes it as fact that our way to conceive of personal identity excludes the possibility of speaking Contacti. He even grants that it is factually impossible to stick to a perdurantistic view of personal persistence as defended by Parfit. (Hirsch 1982, 311) But for Hirsch these are exclusively psychological facts; there are no ontological facts causing us to think in a particular way on persistence.

I have hinted at some of the consequences of a conventionalistic account of persistence. One strategy to avoid them consists in opting for a richer ontology allowing more ontological facts, viz. an ontology with endurers.

Reasons for opting for endurers

The decisive reasons for postulating endurers and thus diachronic identity are neither scientific nor do they result from mere theoretic philosophy. They are due to our *subjective experience* of being intentional agents programming our future and memorizing our past. The mere theoretic arguments in the debate between endurantism and perdurantism seem to balance each other. But the presuppositions of practical rationality, agency, subjectivity, indexicality, and so on, support endurantism.

By acting and feeling, we presuppose that others, and that objects we deal with, go with us in the course of time. We treat them as endurers and are convinced that that is the condition for getting acquainted with them. We conceive of and remember objects as lasting in time. This experience strengthens our every-day conception that not only we as agents having first person access to ourselves endure, but objects endure as well. We are firmly convinced that they continue to exist at least as long as we keep on perceiving them. In real life, we do not construct objects in their temporal continuity out of momentary stages, but because of the mentioned experience presupposing that we can observe how they go on existing.

If it is possible – as we have seen – to interpret the application of the 4-dimensional space-time system as compatible with an ontology with continuants, on the whole, we have good reasons to accept diachronic identity. The fact that objects of empirical scientific research, the objects of natural science, are considered to be four-dimensional, does not imply that they are extended in time too.

Qualitative and spatio-temporal continuity conditions alone are neither sufficient nor necessary for tracing one and the same object through time.

I shall try to argue for the thesis that the individual sortal, the individual sortal determination, guarantees identity through time.

Sortals

With sortal-expressions we state *what* things are. When we are asked what a given thing or animal is, we answer with a sortal-expression. Once a speaker understands the meaning of a sortal expression, she is normally in the position to identify the different instances of the sortal in question. This is also applicable to the temporal aspect of the individuals, as sortal expressions convey the continuity/persistent-conditions and thus in a certain sense existence conditions in time too.

Sortal expressions viz. concepts delimit what falls under them in a definite way and do not permit any arbitrary division of them into parts. A part of a cat might be an organ of a cat, but not a cat. If one grasps the sense of a sortal expression, one is able to count the individuals falling under them.

Sortal expressions play a crucial role in realistic as well as in conventionalistic philosophies. In conventionalism the rules of the use of these expressions and the continuity conditions linked to them are taken to be conventional. In realistic philosophies, based on ontologies with endurers, the persistence conditions are 'given'.

Quine gives a radically conventionalist account of the reference of sortal-expression viz. count terms. The reference of such terms is empirically underdetermined. On the basis of alternative identity-criteria we would count rabbits differently. In such cases we could have far more rabbits in our world then we do. A radical conventionalist account trivializes the question of the persistence of living beings.

There are limits to the conventionalist freedom of inventing sortals with alternative identity and continuity conditions. Indeed, if we could invent sortal concepts simply at will, 'then the real content of the assertion that something lasted till t and then ceased to exist would be trivialized completely.' (Wiggins 2001, 65) We can neither prolong nor abbreviate the duration of the existence of continuants merely by applying different sortal concepts.

If, on the other hand, the range of possible answers to the question *what is x?* is limited, we have to state what makes the difference between *genuine sortal expressions* referring to the basic sortal constitution of an object and the countless expressions which only look like sortal expressions.

A genuine sortal must, at any rate, be relevant for determining when a continuant falling under it really ceases to exist. It allows not only to single the object out and distinguish it from other objects of the same kind, but *to trace* it through time as well. It allows us to determine – at least to a certain extent – what can and cannot befall it, and thus determine what changes it can tolerate without ceasing to exist. If this is so, one has to distinguish between the stuff making up a substance and the sortally determined thing or living being itself.

Objects are not identical with the material they are made of or with the sum of their parts. Their material does not exclusively determine them. We have to distinguish between sortally determined objects and parcels of matter. The latter are not objects in a strict sense. Even where they cohabit, the ways in which they persist through time differ widely. (Runggaldier 1998, 366f.) What e.g. destroys one may entail only a slight alteration in the other. Objects can go out of existence without their parts ceasing to exist, and on the other hand their parts may go out of existence without thereby destroying the objects they constitute.

The sameness of material composition is not necessary for the identity through time of the objects it makes up. Organisms that renew their cells and artefacts are often altered gradually in material parts. Retention of the material composition or sameness of material is on the other hand not sufficient for the identity of an object, even though it is for chunks of matter. Melting down a statue destroys it despite the fact that its material composition remains the same.

The kind of composition, the mode of activity, their way to act and react, their basic capacities, their dispositional properties and functions are relevant for the continuity conditions of sortally determined objects.

Statements on natural kinds are not universal quantifications

What is the status of general statements on living beings of a given kind? What are we referring to when we claim that tigers are carnivorous and wild? Empiricists and conventionalists tend to account for statements on sortals viz. natural kind as universal quantifications, i.e. on the individuals of a set.

A basic problem for a radical conventionalist account of such statements stems from the willingness to accept *exceptions*. Biologists and people generally assume that general statements on natural kinds can be valid even though not all the individuals of the kind satisfy them. We certainly stick to the truth of the claim that cats are four-legged and at the same time believe that many cats live their lives with only tree legs.

General statements on natural kinds in biology or in every-day life differ from general statements in physics, since they allow for exceptions. General statements in physics are law-like and are, thus taken as generally valid. This is not so in the case of general statements on species or their individuals. We take these statements to be valid for those individuals of a species that are normal or healthy.

Statements like 'cats open their eyes on the sixth day after birth' cannot be taken to state a universally quantified truth. They refer to the normal development of an average cat. The reason for the opening of the eyes is not a certain combination of antecedent conditions necessarily leading to it, but rather the nature of a normal cat. Of course the opening of the eyes presupposes certain conditions, but these are not like those antecedent conditions allowing predictions in physics. Even if all conditions for a baby-cat are their best, it might still happen that the cat does not develop like a normal individual does.

To claim this does not imply that there are no statements on living beings functioning as general statements in physics. The general statements on species in biology and every-day life allow, however, for the tricky distinction between normal and abnormal, between healthy and unhealthy, between function and dysfunction. They express regularities in the normal development of an individual of a species.

The regularities referred to are like norms compatible with deviations. Violations do not abolish the norm. General statements on species say how an individual should develop and behave if it is to be regarded as a normal individual of its kind. It can, however, belong to the kind, even if it does not develop like that. There is a logical possibility for various deformities.

We might see these norms as analogue to industrial norms stating how a machine of a certain type has to look like and how it has to function. Of course, it only is an analogy, for in the case of artefacts we look for the thing or person that is responsible viz. guilty if an individual is malfunctioning or has not been produced according to the industrial norm. In nature, we do not look for the guilty. However, we have standards or norms allowing us to distinguish not only between different kinds but also for distinguishing between healthy and well functioning individuals from malfunctioning, unhealthy or deformed individuals. Even Millikan grants that she made use of normative notions by introducing the technical term 'proper function': 'Normative terms are used to indicate any kind of measure from which actual departures are possible… 'With that kind of sky in the west it ought to be sunny tomorrow.' (Millikan 2002, 116)

It is irritating to speak of norms in the realm of nature, since norms in a strict sense fall exclusively into the realm of human action. However, the statements on natural kinds are not merely descriptive, even though they convey empirical knowledge on a natural kind gained in the past.

The persistent and continuity conditions of a living being depend on the kind or sortal of the individual. This is, thus, relevant for how the living being develops during the span of its life. Decisive for it are the capacities or dispositions of its kind. This, however, does not imply – as we have seen – that all individuals of a kind develop in the same way. Some might degenerate but nonetheless remain members of the kind. The members of a species can loose various capacities and gain new hitherto unknown capacities. The basic or essential capacities are, however, relevant for their persistence and continuity conditions.

Capacities and powers

If sortal continuity is guaranteed by the continuity of the essential capacities, then these must be real. Empiricists and conventionalists, however, have to explain them away.

Focussing on chemistry makes it easier to stick to an account for capacities and natural kinds in realistic terms. On closer examination, the questions of identity of chemical elements and compounds do not depend on conventions, or on our interests, psychologies, languages, practices or choices. Chemical substances differ from one another in their internal constitutions. (See e.g. Ellis 2002) Chemical substances behave as they do not only because of external stimuli, but because of intrinsic conditions too. On the exterior side we often have mere triggering conditions, whereas on the internal side, powers, capacities, enabling conditions or – in Dretske's terminology – structural causes.

Harré hints at two distinct paradigmatic cases for tackling these problems. (Harré and Madden 1975, 82ff) One is the mechanical model, exemplified in the forces acting from the exterior between billiard balls. It is sufficient for accounting for mechanical changes. The other model is that of the activity of a living being originating from within, or the chemical reactions between chemical elements, or the explosion of dynamite. If we want to account for all kinds of changes in nature, we can't restrict ourselves to the first model.

Tough empiricistically minded, Nancy Cartwright defends the reality of powers viz. capacities (Cartwright 1989): Capacities are a part of the scientific image of the world, and cannot be eliminated. They are not mysterious, for they can be measured, and scientists measure capacities. Even physical laws such as the law of gravity or the laws of electromagnetic repulsion and attraction can and should be interpreted as laws about enduring capacities. This position is not generally opposed to empiricism, even though it is in opposition to Hume's account of causation as regularity or to the mainstream probabilistic accounts of causality. (Cartwright 1989, 3)

The empiricist and conventionalist deflationary accounts of causal powers have this disadvantage due to all accounts of causality based on mere *regularity relations*. Within this tradition it is difficult to distinguish between mere accidental regularities and regularities due to individuals or substances involved.

One of the main problems for a thorough regularity account of causal laws is due to causal interactions, especially if we look at chemistry: An acid and a base neutralize each other. The specific chemical effects of both are eliminated. (Cartwright 1989, 163) When an acid and a base mix, their effects are not combined with each other; neither can operate to produce any effect at all. The generalisations in sciences like chemistry therefore need further specifications by the condition '*as long as there is no interfering causal interaction*'. Cartwright asks how this can be done without admitting interactions into the descriptive content of the world. (Cartwright 1989, 163f)

An interaction is a special kind of process with an identity of its own. The positivistic attempt at explaining away interactions trivializes this intuition. In order to explain away capacities, one needs some separate, independent characterization of interaction, employing no further concepts like the concept of capacity itself. But this does not seem to be possible.

Let us think of an example: will the rain this holiday weekend result in more or in fewer road accidents? It is not easy to answer the question because the rain has different, opposing effects: by keeping people at home, it tends to prevent accidents; by worsening the conditions of the roads, it will tend to produce them. How should one try to solve the problem solely probabilistically?

Functions and teleology

In biology and in every every-day life, we speak of functions of organs or features of organisms as dispositions to contributions to a certain goal or output. Nobody would e.g. deny that human kidneys perform the function of excretion or that the function of a heart is to pump blood. If one aspires to an explanation of the presence of certain organs or materials, one looks at what they are for.

Talk of functions and teleology in nature is, however, very controversial. Empiricists and conventionalists relegate such talk to the practical domain of human action. They presuppose that teleological explanations have, at most, a heuristic role: they neither have any explanatory value, nor any ontological bearing.

Nowadays, we have a renewal of teleology due to biology. Biologists assume functions and use teleological language, expressions such as 'flee from…', 'protect for the sake of…', 'hide from…' and 'migrate to…'. To ascribe goal oriented behaviour to organisms is, however, not yet to ascribe intentions in the human sense. Many presuppose that the purposiveness of part of the behaviour of living organisms be analysable in some non-mental way.

There are basically two different types of teleological talk: *agency-centred teleology* and *teleology pertaining to natural organisms* (biological processes e.g. occur for the sake of self-preservation or preservation of species viz. parts of the organisms are present for the sake of the organism possessing them).

Some contemporary realists assume teleology even for chemical substances. In his fivefold characterization of *powers,* Molnar e.g. states that they are intentional. (Molnar 2003, 64)

It is important to note that, whatever position one is inclined to defend, realists see this natural teleology or directedness towards a goal as non-purposive, non-mental, non-rational. It is moreover misleading to reject Aristotle's teleology on the grounds of an alleged reading purposive behaviour into natural events.

Ariew sees Aristotle's arguments for teleology mainly directed against materialism, against the thesis that materials and their causes are sufficient to explain all physical events. They are suited for the explanation of some events or processes, but not all.

We have seen that one of the central problems for a conventionalist account of statements on species or natural kinds is due to the exceptions in nature. We distinguish between normal, healthy developments and violations of the norm. In order to do this, we recognize patterns of arrangement and sequential order in nature – like the patterns in nest building of birds and web making of spiders and so forth. Deformities are due to aberrations from these patterns. These occur when one feature, organ or part of an organism does not perform the function it should perform. Orderliness among functional relationships is a presupposition for normal developments in nature, whereas disorder leads to aberrations, even monstrosities. Ariew points out that according to Aristotle, the materialist cannot explain what goes wrong when mistakes occur or what goes right when developed things work. (Ariew 2002, 17)

It certainly is a mistake to think that Aristotelian teleology presupposes conscious intentionality or bizarre backwards causation. Ariew concludes thus: 'If biology has an ineliminable teleology, this is not so bad as long as it is one of the more restrained Aristotelian versions of teleology.' Interestingly enough, Aristotle uses for the actualised sortal determination interchangeably the two terms 'ἐνέργεια' and 'ἐντελέχεια', the second term having the connotation of the achievement of a τέλος: as soon as a living being has actualized itself, it has reached its τέλος.

If we follow these Aristotelian intuitions on goal oriented processes, we see that the assignment of functions should not be limited to purposive conscious human behaviour. Goal-directedness is not necessarily a mental property. As an objective property it can be assigned to other living beings as well. Thus, function statements do not only have a heuristic value, useful for practical purposes, they can be true in a literal sense as well. They can be statements on the contribution of traits, parts or processes to the achievement of a goal.

There are various ways to unfold these realistic intuitions. One way is centred on the 'causal-role analysis' (conf. Cummins 1975), the other more on the 'etiological analysis' (conf. Wright 1973, conf. also Millikan 1984). For my purposes it suffices to hint at the realistic interpretation of some of our functional talk. An item's function is an objective dispositional property to causally contribute to a goal or an output capacity of a complex system. If the realist Aristotelian intuitions are justified, it is obvious to assume that the continuity conditions of individuals depend on their fundamental functions too.

To sum up: the persistence conditions of an individual depend on its individual sortal or essence. There is a mutual dependence relation between the continuity of the sortal and that of the basic capacities, powers and functions. These should not be reduced to ways of speaking with a mere heuristic role, but have an ontological bearing.

References

Ariew, A. (2002) 'Platonic and Aristotelian Roots of Teleological Arguments' in: A. Ariew, R. Cummins, M. Perlman (eds.), *Functions*, Oxford: Oxford University Press, 7-32.

Aristotle, *Metaphysics*.

Carnap, R. (1928) *Der logische Aufbau der Welt*, Wien: Felix Meiner Verlag.

Cartwright, N. (1989) *Nature's Capacities and their Measurement*, Oxford: Clarendon Press.

Cockburne, D. (1998) 'Tense and Emotion' in R. LePoidevin (ed.), *Questions of Time and Tense*, Oxford: Clarendon Press, 77-91.

Cummins, R. (1975) 'Functional Analysis', *Journal of Philosophy* 72/20, 741-765.

Denkel, A. (1996) *Object and Property*, Cambridge: Cambridge University Press.

Ellis, B. (2002) *The Philosophy of Nature*. Chesham.

Gale, R. M. (2002) 'Time, Temporality, and Paradox' in R. M. Gale (ed.), *The Blackwell Guide to Metaphysics*, Oxford: Blackwell, 66-86.

Harré R. and Madden E. M. (1975) *Causal Powers*. Oxford: Blackwell.

Hirsch, E. (1982) *The Concept of Identity*, Oxford: Oxford University Press.

Millikan, R. G. (1984) *Language, Thought, and Other Biological Categories: New Foundations for Realism*, Cambridge/Mass.: M.I.T. Press.

Millikan, R. G. (2002) 'Biofunctions: Two Paradigms' in: A. Ariew, R. Cummins, M. Perlman (eds.), *Functions*, Oxford: Oxford University Press, 113-143.

Molnar, G. (2003) *Powers. A Study in Metaphysics.* (ed. by S. Mumford) Oxford: Oxford University Press.

Quine, W. v. O. (1960) *Word and Object*, Cambridge/Mass.: M.I.T. Press.

Quine, W. v. O. (1963) *From a logical point of view*, New York: Harper.

Runggaldier, E. (1984) *Carnap's Early Conventionalism*, Amsterdam: Rodopi.

Runggaldier, E. (1998) 'Sortal continuity of material things', *Erkenntnis* 48, 359-369.

Wiggins, D. (2001) *Sameness and Substance Renewed*. Cambridge: Cambridge University Press.

Wright, L. (1973) 'Functions', *Philosophical Review* 82, 139-168.

POWERS AND PERSISTENCE

STEPHEN MUMFORD

1. Introduction

Recent years have seen a rise and development of the ontology of dispositions or causal powers (for example, see Bird 2007, Ellis 2001, Molnar 2003, Mumford 1998, 2004). This is a metaphysics that accepts real, irreducible dispositions that, among other things, involve greater-than-contingent connections between distinct existences. In this paper I address the issue of persistence, which is one that those committed to a powers ontology have not yet considered in any detail. If the world contains real, irreducible dispositions, what if anything would this tell us about the metaphysics of persistence? Would it help us to decide, for instance, between endurance and perdurance theories? Would it tell us anything about how ordinary particulars persist through time?

I will be offering no argument for the existence of causal powers. I will instead be simply assuming such an ontology to be true and then considering where this leaves the question of persistence. This paper could thus be understood as a discussion of the conditional interrogative: if there are powers, where would that leave the question of persistence? There will be nothing that would persuade directly someone of the plausibility of a dispositional ontology if they did not already see it as plausible.

Although I offer no direct argument for the existence of powers, I ought nevertheless to at least say what powers are and what the key commitments are of a dispositional ontology. I will not do so at any length because my purpose is to address the question of persistence. But we need at least enough to understand how the issues of powers and persistence relate. The idea of a disposition is known to us all through familiarity with cases such as solubility, fragility, elasticity and hardness. These are frequently understood as a species of properties, distinguished by them somehow containing within themselves the possibility of further properties. Hence solubility is understood as a property in its own right but one that can produce a further property, becoming dissolved. In this case, there are well-known circumstances in which a soluble thing will become dissolved, namely immersion in a liquid. When a soluble thing is in liquid and then becomes dissolved, we would say that the disposition has manifested itself or we could say that being dissolved is the manifestation of solubility.

Why are we interested in such properties, which might seem to be a small, unusual, and not very scientifically valid sub-class? Proponents of the dispositional ontology have argued on the contrary that many properties are dispositional in character. The position known as pan-dispositionalism is that all properties are dispositional. As Popper (1959: 420) said, even being dissolved is to have a disposition: for a solid to be recovered upon dehydration. Even to have a certain shape – shape being in the past suggested as a paradigmatic non-dispositional property – is after all to be disposed to behave in a certain way. Furthermore, the fundamental properties with which science deals are among those that look to be dispositional in nature – properties such as spin, charge and mass – so it looks problematic to call dispositions non-scientific.

If there are such properties, what does it mean? One thing that the powers theorists have highlighted is how properties could be understood as a causally interconnected web. Property A would be a power for another property B. But B would be itself a power for some further property C, and so on. As Armstrong has remarked (2005; 314), causality becomes the passing around of powers.[1] A second notable feature is the interdependence of the properties. Dissolvedness is a part of the essence of solubility. There could be no solubility if there was not also a property of being dissolved, and no fragility unless there was such a thing as being broken. While some particular thing can remain fragile without ever being broken, the property of being fragile is real only if the property of being broken is also real. The properties thus become essentially and necessarily connected even though they remain distinct existences and it is this feature that makes such powers abhorrent to traditional Humeans.

2. How do powers relate to persistence?

There are at least three ways that the notion of a power relates to the question of persistence, though only the third way will be pursued here in any detail.

1 Armstrong intended this as a criticism of the powers ontology but instead the powers ontologist could make this a virtue of their account (see Mumford 2009).

First, many powers theorist are committed to what is known as the Eleatic test of reality, which says that to be is to be powerful. Having power is the mark of being real. To persist might then be thought to be a retention of power: either the same power(s) or the acquisition of new powers over time. There would be many problems to be resolved here, however. If the persistence of particulars is explained in terms of persistence of powers then clearly the key notion of persistence would remain unexplained. But even apart from that, one would still need to explore the exact relation between the persistence of particulars and the retention of powers. After all, couldn't a particular persist through a moment *t*, a moment of radical change, at which each and every one of its powers is lost and replaced?

The second way in which powers and persistence might relate concerns whether persistence itself could be understood as a kind of power. Could there be a power to persist? A problem could be that a power of persistence might look like a *deus ex machina* for persistence, which merely states that there is such a power of persistence as some kind of magical 'solution' to a very real and difficult problem of in what persistence really consists. Nevertheless, there are some considerations that suggest an idea of persistence as a power cannot be immediately ruled out. Clearly my body has a power to persist. At a macroscopic, biological level, it is able to take energy from food and sunlight and process those resources so as to create a later stage of itself, complete with those same self-sustaining powers. And at a microscopic level, particles of matter can be understood as parcels of energy, energy for self-perpetuation. In contrast, some things are not empowered to persist. A soap bubble has a relatively fleeting existence, its surface tension being insufficient to sustain it through the harshness of its environment. Some particles have very short half-lives, seemingly having within themselves some principle for their own destruction. A bridge may collapse or some component fail because it is insufficiently robust for the circumstances. In other words, it does not have a power to persist.

The third issue concerns how stages or temporal parts are connected within the perdurance theory of persistence, and this is the issue on which I will now concentrate. Perdurantism, as standardly understood, is a view that things persist through having a series of discrete and momentary parts or stages that are connected in some appropriate way. Might this appropriate way be explained in terms of causal powers? Or does a powers ontology show that one should instead favour an endurantist ontology, with particulars genuinely enduring through time and 'fully present' at each instant at which they exist, as opposed to only one temporal part being present at each time? Ultimately, it would be satisfying if the powers ontology were able to deliver for us a verdict on the endurance/perdurance debate. I will argue that it doesn't quite do that – not conclusively anyway – but that a powers ontologist nevertheless has some reason to favour endurantism as sitting most comfortably with their ontology.

I will argue that if one is to be a causal powers theorist, one will have difficulty defending a perdurance theory of persistence. This is because perdurance theory has on ontology in which the key entities are unchanging or static and discrete. Where connected together, they are connected by external, non-supervenient relations. The perdurance theorist may try to argue that these relations are simply the everyday causal relations. But external and contingent relations do not make for a satisfactory account of causation by the lights of the powers theory. If one were to get an account of persistence from a theory of powers, therefore, it would be an endurance theory.

3. The metaphysics of perdurance theory

Although I would accept that there is scope for some variety among perdurance theories of persistence, I will take my lead from Katherine Hawley's recent characterization of the view. She says that "Perdurance and stage theories share a common metaphysical picture – the world is full of very short-lived objects existing in succession" (Hawley 2001: 42). I will not go into the details of the difference between a perdurance and a stage theory (for more on that, see Hawley: ch. 2) as it is the metaphysical picture of a series of short-lived existents that is the focus of my discussion.

Perdurance theory is motivated in part by a desire to explain change or, what Lewis (1986a: 202) calls the problem of temporary intrinsics. The same thing cannot bear contraries. So as not to ascribe incompatible properties to the same thing, the theory ascribes them to different temporal parts of the thing. We cannot say that the same thing, *a*, is all red and all green so we should say instead, according to the perdurance theorist, that *a*-at-t_1 is red while *a*-at-t_2 is green. *a*-at-t_1 and *a*-at-t_2 are two different entities – two temporal parts of *a* – and thus to say that one is all red and one is all green is not to ascribe contrary properties to the same thing. Temporal parts (or stages, in stage theory) have to be unchanging, therefore, in order to solve the problem for which they were designed. The theory attempts to explain change in an object in terms of different temporal parts of that object so the temporal parts must themselves be unchanging if the explanation is to be all-encompassing.[2]

Such temporal parts are not just changeless, in Hawley's version of the theory, but they are also fleeting existents. She considers the question of whether the temporal parts last as long as the actual changes in an object or only as long as its possible changes. We can at least assume that there is an object that undergoes no change at all for an hour, for instance, so why not say that it has an unchanging temporal part of one hour's duration? But Hawley (2001: 49) sees a reason to say that the temporal parts of an object are as fine-grained as possible changes. An object that does not change through time nevertheless could have changed. And how could it have changed unless it had some distinct temporal parts that could have instantiated different properties to the ones they actually instantiated? An object, then, does not have merely as many temporal parts as actual changes but as many temporal parts as possible changes. Temporal parts are thus very short lived indeed. Perhaps there are as many temporal parts of a thing as there are times at which that thing exists. Temporal parts would then be as fine-grained as time itself.

To account for change, therefore, the perdurance theory posits for each thing, a large number of changeless and fleeting temporal parts of that thing. They will also be entirely distinct and discrete from each other, permitting no overlap.

It should be noted before moving on, however, that the main alternatives to perdurance theory – endurance theory and adverbialism – also offer explanations of change that answer the problem of temporary intrinsics. The endurance theory would index properties to times. Hence instead of saying that something is both all red and all green, where it has been subject to change the account would say that a is red-at-t_1 and a is green-at-t_2, which involves no contradiction. The adverbialist prefers to say that a is-at-t_1 red and a is-at-t_2 green. There are thus at least these three ways to account for the temporariness of intrinsic properties. Our decision between endurantism, perdurantism and adverbialism does not, therefore, come down to one theory being able to account for change and the others not. Instead, we would have to consider other features of each of the theories. Hawley, for instance, makes much of the general metaphysical appeal of four-dimensionalism. We are very familiar and accustomed to understanding things as having spatial parts. Time is now understood to be analogous to space. If things have spatial parts, therefore, we should expect them to have temporal parts.

2 It may be possible to develop an ontology in which there are temporal parts that do contain change. Whitehead (1929) can be interpreted in this vein. Clearly this version or perdurantism would have to say something else about the problem of temporary intrinsics.

While I accept that perdurantism has this kind of appeal, I will argue that it also has a draw back. While it solves one problem of change, I argue that it leaves another.

4. The metaphysics of powers

Powers theorists believe in a world containing continuous, dynamic, natural processes involving active particulars. Earlier portions of such processes produce the later portions in a continuing and continuous process. Dispositions are tendencies towards such natural processes, as when a soluble thing dissolves, a magnetic thing attracts, a flexible thing bends and a fragile thing shatters.

Ellis (2001) is a dispositionalist who argues that there are natural kinds of processes, which have their properties or stages essentially. A kind of process P, such as photosynthesis, would not be the process it was unless it involved such and such features, for instance, being a botanical process of energy being taken from sunlight. And there are some natural kinds of thing, such as plants, that would not be the things they are unless they partook in the P-kind of process.

Dispositionalists think that the dispositionality in such processes is real and irreducible. Plants tend to photosynthesize. This does not necessitate that they do so, as they might be kept out of the light or an individual plant may be infected with a virus that countervails its photosynthesizing tendency. Having a natural disposition does not mean that it has to be manifested, therefore. The process is essentially dispositional in nature and cannot be cashed out in other terms.

The world's particulars are seen as active in that they have within themselves the principle of change. If a particular has a disposition then it intrinsically tends towards some kinds of process rather than others. Its behaviour in this respect is not determined externally or extrinsically, with it being a passive recipient of change. It is disposed to behave in a certain way as part of its own nature. Such particulars are also seen as dynamic, as acting particulars (perhaps *always* acting).

This leads us to this issue of continuity. Change in a particular can be understood as a continuous and constant process in the sense that it contains no proper portion that is not undergoing change. While the whole process involves movement from one state, condition or property to another, each proper portion of that process also involves change. The process should not, therefore, be broken down into its static, instantaneous parts. If one tried to understand the process in that way, one would lose something of its dynamic and developing nature. To understand this, we need to consider what the perdurantist would make of processes.

5. Process and perdurance

To account for processes, the perdurance theorist can only offer a sequence of parts or stages, each of which is static. Such parts must be suitably related if they are to genuinely form a sequence. Not any old collection of temporal parts forms a sequence. They must be suitably related so that they are temporal parts of the same thing. Changes in the same thing could then constitute what I am calling a sequence or process. What kind of relation must the parts bear to each other to be a sequence? Without saying exactly what the appropriate relation is, for perdurantists to posit in order to stick stages together, Hawley does tell us that it is a non-supervenient relation. This is because she cannot see how facts about processes of a thing could supervene on the facts about the static properties of the changeless temporal parts of that process. A non-supervenient relation is thus an external relation: one that does not hold merely in virtue of its relata existing. The holding of an external relation is thus some further fact, over and above the intrinsic facts about the relata.[3]

It makes sense to us to allow that there could be a homogeneous spinning disk. This possibility creates a problem for perdurantism because each static temporal part of this disc's spinning will look like all its other temporal parts. The intrinsic, changeless properties will look the same for each part of the process. Among other things, this threatens Lewis's doctrine of Humean supervenience (Lewis 1986b: ix-x). Hawley thinks this shows that one has to allow that the relation that sticks temporal parts or stages together is non-supervenient; that is, it is not determined solely by the parts and their intrinsic properties (Hawley 2001: 73). The persisting disc has a property – it is spinning – that is not determined solely by the properties of its temporal parts. Hawley then generalizes this conclusion: the relations that hold between distinct temporal parts or stages, that make them parts or stages of the same thing, are not determined solely by the intrinsic properties of those parts or stages. The holding of the appropriate relation is thus some further fact in the world, above and beyond the facts about the individual parts.

What then can this type of perdurantist say about processes? A process is essentially extended over time, which for perdurantists means extended over many temporal parts. It must thereby consist in a series of

3 External relations contrast with internal relation, which exist whenever their relata exist.

temporal parts, suitably related by the appropriate non-supervenient relation, where at least one static property of each temporal part differs from the properties of a neighbouring temporal part. Such a perdurantist process is illustrated in figure 1. S1-S3 are stages or parts and properties F, G and H are all static properties, where a static property is a property capable of being fully instantiated at an instant and thus involving no change.

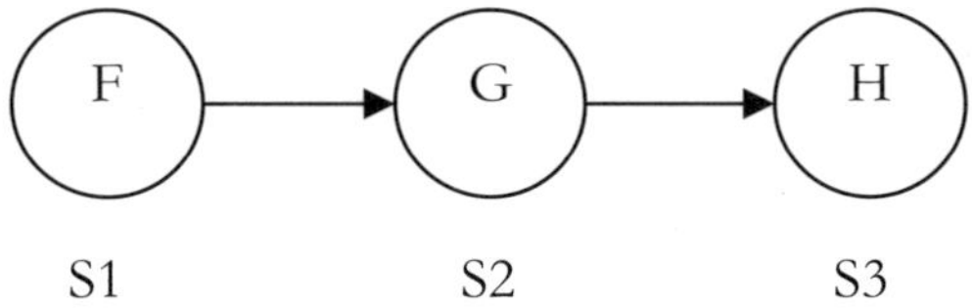

Figure 1

This immediately creates a problem of accommodating dynamic properties: those properties that are essentially changes and thus cannot be instantiated at an instant. If a particular is heating, rotting, growing, and so on, then it has a dynamic property. The perdurantist cannot permit such essentially dynamic properties as real and irreducible. Instead, they have to be reduced to sequences of static properties. Hawley does this for what she calls 'lingering' predicates, such as the purring of a cat. It may seem an unnecessary linguistification of the issue to talk of predicates rather than properties but we can see the perdurantist strategy nevertheless. As she says, 'lingering' predicates can be satisfied by a stage when it 'is suitably surrounded by and related to other stages with appropriate properties.' (2001: 54). Here, then, is something that will immediately be unattractive to the powers theorist. Only static properties can be taken as primarily real. Dynamic properties and processes have to be constructed from sequences of static properties suitably related. The powers theorist is likely to see this emphasis as wrong. Dynamic powers and processes are fundamentally real in the dispositional ontology.

This is not all the powers theorist is likely to find objectionable. If the relations between S1 and S3 are non-supervenient, as Hawley insists, there is nothing about the mere existence of S1-S3 that makes it that they are indeed so related. We have seen that the relation is an external one. S1-S3 could, therefore, exist without being suitably related, which means that it is a contingent matter that they are so related. There is, therefore, nothing more than contingency in this process. There is nothing that makes S3 instantiate property H, even if the stages S1 and S2 of the same particular thing have instantiated F and G. The dispositionalist, in contrast, believes in real dispositions towards natural kinds of process. If a particular instantiates G, having previously instantiated F, it may well be naturally and essentially disposed towards H.[4] In contrast, just instantiating G, having not instantiated F immediately before, might leave that particular not naturally disposed towards H.

6. Dynamic powers

Could the perdurance theorist combine their ontology with the powers ontology so that there is the requisite connection between stages? It is hard to see how they could do so while the stages or parts remained static and unchanging, and while processes are constructed using non-supervenient relations.

What I am trying to envisage is a perdurantist ontology in which the requisite non-supervenient relations are provided by the real causal powers of dispositionalism. Each static temporal part or stage might have an added power towards the next part or stage. But there are two reasons why the dispositionalist would find this mixed view unattractive or unconvincing. The first follows from the account of process given at the end of the last section.

Merely instantiating property G should not be enough to dispose towards H if parts or stages are instantaneous and have only static properties. For the dispositionalist, the reality of the natural process is crucial and is what would give one stage a disposition towards another. It is being part of a process that involves G coming from F that disposes stage S2 towards H. Natural kinds of process will have this sequence – F, G, H – essentially. The perdurantist will not have the requisite necessity in their processes. Powers would be attached to single stages and consequently would have none of the constraints brought by the dispositional essentialist commitment. It would be entirely contingent what static properties went with what powers. But a dispositionalist does not see the relation between properties and powers as contingent. A property always disposes its particular towards some natural kinds of process rather than others, and it is necessary that it does so (Shoemaker 1980). A mixed ontology of

4 There are good reasons to think of dispositionality as involving a distinct and irreducible form of modality that is neither complete contingency, nor complete necessity, but something in between. The connection between a disposition and its manifestation is not a completely contingent one, as the disposition tends towards a particular manifestation and sometimes succeeds in causing it. But as any such disposition can be prevented from manifesting itself, such dispositionality is short of necessity. The modality of dispositionality may well then have to be taken as primitive (Anjum and Mumford, forthcoming).

perdurantism, with dispositions added on top, could not secure this crucial feature of dispositionalism.

There is, however, a second reason why the dynamic powers of dispositionalism could not be captured by the static properties view. This is, however, an a posteriori argument concerning the way the world works, and where the a posteriori arguments may not yet be conclusive. The question is how could S1 being F causally produce S2 being G if S1 is a static part or stage? There is nothing going on inside the static S1 that could be the basis of the power towards G. It statically instantiates the static property F. Can it really instantiate powers if that is the case? To do so, there would have to be 'static powers', which could be instantiated at an unchanging instant. But these seem unfathomable and cannot account for all the powers in all the world's processes.

There could well be static powers in the sense that an unchanging, static particular could nevertheless be involved in a change. Fragility could be a case in point. An object could sit, motionless and unchanging yet nevertheless be able to be broken. But this is a case of a so-called passive power, which is a power to receive a change rather than a power to initiate a change. This could contrast with the power of a clock to ring an alarm at a pre-set time. This power is dynamic or active in that the clock has the power to initiate a change, to ring an alarm, without needing anything further to be done to it. The way we think the clock can do this is that it is undergoing a process – such as its short arm being moved towards a trigger – that will in the end produce the alarm. We would, however, be mystified by the idea of something having an active power that had no grounding in that particular's movements or changes. The only cases we can think of are spontaneously manifested powers, such as radioactive decay or the spurious case of spontaneous human combustion. But could we countenance a power that was manifested in a spontaneous but reliable and regular way? It seems hard to think why we would or should and this seems to offer us some insight into the difference between active and passive powers. The distinction is old though seldom explicated in detail. My suggestion here is that a passive power is one that could be instantiated by an unchanging, motionless particular whereas an active power could not be. Consequently, a perdurantist picture would have difficulty accommodating active powers in a credible way.

The world of active processes is the one with which we are most familiar. Consider the case of a force, which I take to be an example of a power. A billiard ball rolls across a table. Its force drives it forward, though it is gradually lost through friction. This is a process through time. In the perdurance view, there are no moving parts or stages, only a series of suitably connected static parts or stages. Presumably, then, there is no momentum in any of the static parts. A scientist may well be able to construct a model in which an unmoving billiard ball can be ascribed a force and momentum, but can we as metaphysicians accept that this is the way the world actually works? If we accepted a hypothesis that there are active powers, which in being exercised and merely in being held involve change, movement and dynamism, then perdurance theory cannot seriously accommodate them. The perdurance theorist may, like the aforementioned scientists, be able to construct a model to explain how powers could be ascribed to successions of temporal parts. But this would be like Humeans saying that their metaphysic contains causation. The realist objects that this is not genuine causation. As Lewis makes clear, in his view the causal relations of a world supervene on the Humean mosaic of discrete, unconnected events. This is not real causation as far as the powers theorist is concerned and they will similarly say that the perdurantist's processes and powers are not real processes and not real dynamic powers.

7. Perdurance and Humeanism

This last issue seems to be the crux. For the perdurance theorist, the distribution of powers in the world will supervene on the distribution of static temporal parts. The powers theorist sees things the other way round: the distribution of properties in the world will be determined by the powers of particulars, together with their history. It is not all contingent what follows from what, while the perdurance theory seems to make it so. It is not contingent what powers attach to which properties and it is not entirely contingent what stages of a process follow other stages. The perdurantist picture seems to have no way of ruling out these contingencies. As argued above, because temporal parts are instantaneous and static, they cannot contain active, dynamic powers of the kind described.

Because of this, it seems that perdurance theory is a variety of Humeanism – a variety that the powers theorist rejects. Those who oppose perdurance theory tend to be those who believe in causal powers, though I haven't yet found this given as a reason to reject perdurantism. Hawley presents in her book two lists, one of perdurantists and one of endurantists. Both lists contain eminent philosophers. Among the perdurantists she lists (2001: 10n) Lewis, Heller, Jubien, Armstrong, Le Poidevin, Noonan, Quine, and Robinson. Among the endurantists, she includes Merricks, Gallois, van Inwagen, Haslanger, Lowe, Mellor, Oderberg, Olson, Wiggins, Thomson, and Simons. This may be pure coincidence but there is also a trend of the perdurantists to be more generally attracted towards Humeanism and of the endurantists to be prepared to take real causal powers seriously. Even if, as a matter of fact, it is pure coincidence that the perdurantists and endurantists take these views of Humeanism and anti-Humeanism, I am suggesting that if one considers the issue of how powers relate to persistence, Humeans and anti-Humeans ought indeed to line up in these ways.

7. Four-dimensionalism

There remains a problem. Four-dimensionalism has its attraction as a view of the totality of the world. All times and facts are equally real. But perdurance theory is often equated with four-dimensionalism (e.g. in Sider 2001), as if they are one and the same. This would leave a number of choices:

i. Reject four-dimensionalism
ii. Reject the causal powers ontology
iii. Produce a perdurance-friendly theory of powers
iv. Work to show that enough of what is attractive about four-dimensionalism can be retained while the doctrine of static temporal parts is rejected.

I am assuming the powers ontology within this paper so I will not here entertain taking option ii. And I have argued during this paper that option iii is not really workable. That leaves options i and iv. I would prefer the latter option. This would thus be a view that, contrary to the way Sider presents the position, nevertheless rejects perdurantism while retaining a four-dimensional view of the world. I have some optimism that this can be done. Accepting the reality of the four-dimensional manifold is one thing but committing to that manifold being populated by fleeting existents with their static properties is another. Instead, the four-dimensional world, in which all places and times exist and are equally real, could be populated by particulars and their dynamic properties, with natural powers and processes doing the work of producing change. Atemporally, there may well be no change in the four-dimensional manifold, but there is change when one considers the temporal perspective – the temporal facts – within that world. The powers ontology, in my view, paints a more convincing ontological picture of how that change occurs than does the perdurantist alternative.

8. Summary

While I accept that this discussion may still fall short of a conclusive demonstration, I argue that a realist about powers should, more naturally, commit to an endurantist view of persistence.[5] In this view, particulars persist by being 'wholly present' at all the times when they exist, rather than there existing only one of their temporal parts or stages at each different instant. The reason a dispositionalist should want to side with endurantism is that the main alternative, perdurantism, has difficulty producing an account of powers and processes that the dispositionalist will find convincing. There are, in summary, at least these three problems in combining a powers ontology with perdurance theory:

1. As each part is static, how can it bear dispositional properties that are essentially dynamic?
2. Processes occur over many stages. How can there be something in one static stage that would determine all the further stages of a process
3. How can there be the requisite production, of a later stage by an earlier stage, that the powers theorist requires?

Dispositionalism and endurantism are, therefore, the more natural bedfellows. [6]

References

Anjum, R. L. and Mumford, S. (forthcoming) 'Dispositional Modality', XXI. Deutscher Kongress für Philosophie.
Armstrong, D. M. (2005) 'Four Disputes About Properties', *Synthese* 114, 309-320.
Bird, A. (2007) *Nature's Metaphysics*, Oxford: Oxford University Press
Ellis, B. (2001) *Scientific Essentialism*, Cambridge: Cambridge University Press.
Hawley, K. (2001) *How Things Persist*, Oxford: Oxford University Press.
Lewis, D. (1986a) *On The Plurality of Worlds*, Oxford: Blackwell.
Lewis, D. (1986b) *Philosophical Papers* II, Oxford: Oxford University Press.

5 The position would also leave the way open for an adverbialist account of change, which also denies that particular persist by having a series of temporal parts.

6 My thanks to Charlotte Matheson for extensive discussion of these issues, as well as Katherine Hawley and those who commented when I presented the paper at the Humboldt's *Unity and Persistence* conference.

Molnar, G. (2003) *Powers*, S. Mumford (ed.), Oxford: Oxford University Press.

Mumford, S. (1998) *Dispositions*, Oxford: Oxford University Press.

Mumford, S. (2004) *Laws in Nature*, London: Routledge.

Mumford, S. (2009) 'Passing Powers Around', *The Monist* 92, 94-111.

Popper, K. (1959) *The Logic of Scientific Discovery*, rev. edn 1980, London: Hutchinson.

Shoemaker, S. (1980) 'Causality and properties' in *Identity, Cause and Mind*, expanded edn, Oxford: Oxford University Press, 206-233.

Sider, T. (2001) *Four Dimensionalism*, Oxford: Oxford University Press.

Whitehead, A. N. (1929) *Process and Reality: an Essay in Cosmology*, corrected edn, D. R. Griffin and D. W. Sherburne (eds.), New York: Free Press.

NATURAL INDIVIDUALS AND INTRINSIC PROPERTIES

GODEHARD BRÜNTRUP

Abstract

In the world there are concrete particulars that exhibit the kind of substantial unity that allows them to be called substances or "natural individuals," as opposed to artifacts or mere conglomerates. Persons, animals, and possibly the most fundamental physical simples are all natural individuals. What gives these entities the ontological status of a substantial unity? Arguments from the philosophy of mind and arguments from general metaphysics show that physical properties alone cannot account for substantial unity. The ultimate intrinsic properties of natural individuals resemble phenomenal mental properties rather than any other kind of known properties. Pan(proto-)psychism is thus supported by systematically related arguments from different areas of philosophical inquiry. A certain amount of skepticism regarding the full nature of absolutely intrinsic properties is nevertheless well-advised.

1. Introduction

In recent analytic philosophy, a substance has mainly been conceived as that which endures through time. The debate between endurantists and perdurantists became thus the centerpiece of many discussions. There is, however, another historically quite influential aspect of substantiality. A substance is something that exists (or possibly can exist) by itself. Substances are ontologically self-sufficient; they do have an intrinsic nature that is not bestowed on them by entering certain accidental relations. This notion of a substance would encompass also very short-lived and event-like entities. If they are part of the ontological array of concrete entities, they are also substances in this sense. Because the notion of a substance is heavily loaded with preconceptions, I choose to call them "natural individuals." They are "natural" because they are to be distinguished from artifacts like computers or TV sets. They are individuals because they are to be distinguished from mere conglomerates, like a pile of stones or a cloud formation. I wish to argue that there is a substantive theoretical link between the discussion about the intrinsic natures of natural individuals and some developments in the philosophy of mind.

In recent philosophy of mind, the discussion of the "hard problem of consciousness" has resulted in a widespread acceptance of the thesis that the qualitative mental properties of phenomenal experience cannot be fully reduced to physical properties. It is the intrinsic qualitative aspect of the mental that cannot be captured by the causal and functional concepts of the physical sciences. Because of their intrinsic nature, qualitative mental states cannot be captured by scientific analysis. The properties that science finds are all dispositional. Dispositions require a categorical (non-dispositional) basis on which they supervene. For present purposes I will mean by "intrinsic properties" those properties that a thing has in itself, independently of its relations to other things: the properties it could have even if it were the only thing in the universe. The mental properties of the Cartesian thinking thing are intrinsic in this sense. It can be conceived that all of its mental states could exist even without a material external world. The richness of its intrinsic properties is in principle independent of external relations (maybe with the exception of God). Thus, in contemporary philosophy of mind, functionalism has been criticized as capturing only the relational aspects of mind, and missing the intrinsic, qualitative mental properties. Can physical properties be intrinsic in this sense? That is a difficult question. Mass is by some considered to be an intrinsic property. But then, having a mass of m is a property such that something that has that property will play a certain functional role defined by a relation of force and acceleration: $m=F/a$. What about rest mass? It is a consequence of general relativity that only an isolated system would have a coordinate-independent mass. Since a non-isolated system is constantly exchanging energy-momentum with its environment, the mass at a certain point in time would depend on the simultaneity determinations of the observer. In quantum mechanics mass is ultimately explained by the Higgs mechanism, which clearly is a functional concept. It seems that, if only we dig deep enough, even physical concepts that looked *prima facie* like non-relational intrinsic properties, turn out to be defined relationally. It is thus reasonable to ask: What is the intrinsic categorical nature of those entities that are relationally defined by physics?

A similar question was asked in early modern philosophy about the intelligibility of the Cartesian notion of a material substance as an extended thing. The main argument here is that extension is a relation which cannot on its own constitute a substance. It presupposes some intrinsic nature of the relata, the terms of the relation. A system of spatial relationships is too incomplete to constitute a real concrete entity. More recently, Max Newman (Newman 1928) argued against Russell's view that we know only the structural features of the world. He claimed that, unless we take into account the intrinsic features of the relata, there will be too many relations. The existence of a set of relations is trivially true of a set of objects unless the relata have some qualitative intrinsic properties. Mere relations are just sets of ordered sequences of entities. And there are too many of them for any given number of objects, unless the intrinsic nature of those objects determines a specific relational structure. Newman reads this as a reductio against relationalism, claiming that a relationalist view of the world lacks any substantive ontological content except for a trivial claim about the cardinality of individuals in this world. One can also read this idea more in the line of Putnam's "model-theoretic argument" (Putnam 1980). There are too many ontological interpretations (models) of our theories. Our scientific descriptions of the world are unable to single out the intended model, i.e., the *real* world. Similarly, if objects are mere nodes in a relational graph with no intrinsic nature, then too many relational graphs are possible. Since science deals only with the relational structure, not the intrinsic natures, we can never know the one true story about the world in a metaphysical-realist way. We have too many "truths."

It seems that what is missing in the merely relational definition of a physical entity is some intrinsic qualitative aspect that grounds and determines the relations. Possible candidates for such grounding intrinsic properties are (proto-)mental or (proto-)experiential properties. If that is the case, then a realistic physicalism or naturalism entails panpsychism, as has been recently argued by authors like David Chalmers, Galen Strawson and Gregg Rosenberg.

2. Intrinsic properties and the philosophy of mind

Many arguments have been advanced to show that facts about qualia are not implied by physical facts. The most famous being the one about Mary the perfect neuroscientist who has never seen a color. Initially, I will focus here on David Chalmers' famous "zombie argument" because it helps to clearly indicate where in the logical structure of the debate the physicalist is forced to draw panpsychist conclusions. The general form of the argument is this (Chalmers 2002, 249): Let P be the conjunction of all microphysical truths about the universe, and let Q be an arbitrary phenomenal truth about the universe.

(1) P&~Q is conceivable.
(2) If P&~Q is conceivable, P&~Q is metaphysically possible.
(3) If P&~Q is metaphysically possible, materialism is false.
(4) Materialism is false.

But, says the standard scientific essentialist (type-B materialist), P&~Q is only conceivable but not metaphysically possible. The mistake, according to the type-B materialist, is that we are working with primary intensions when dealing with epistemic possibility (conceivability), and with secondary intensions when dealing with metaphysical possibility. In the first case, we consider a world as actual; in the latter case we consider a world as counterfactual (how things might have been but are not). If we consider Putnam's XYZ-world as actual, then "water is not H_2O" is true. This is an instance of an epistemic possibility. If we rigidly hold the meaning of "water" fixed by the actual world and counterfactually consider the XYZ-world, then "water is not H_2O" turns out to be false. Then "water is H_2O" expresses a Kripkean necessary truth. Let us call possibility associated with primary intensions "1-possibility," and possibility associated with secondary intensions "2-possibility." In order for Chalmers' argument to meet the challenge of type B-materialism, it should rather look like this (Chalmers 2009):

(1) P&~Q is conceivable.
(2) If P&~Q is conceivable, P&~Q is 1-possible.
(3) If P&~Q is 1-possible, P&~Q is 2-possible.
(4) If P&~Q is 2-possible, materialism is false.
(5) Materialism is false.

Here, the truth of (3) requires that both P and Q have primary and secondary intensions that coincide. In the case of Q, this seems unproblematic. If something feels like pain, it is pain. If something feels like consciousness, it is consciousness. In the case of P, however, the issue becomes much more problematic. Physical properties are functionally defined. We can say that the primary intension of "mass" picks out whatever plays the mass role in a given world. We can also say that the secondary intension of "mass" is tied to the property playing that role in our world in such a way that in a world where something else plays the mass role, this role filler is not mass. Premise (3) can be rejected on these grounds. But what does that mean? In that case there would be possible worlds that verify the structural-relational description of our world in physical terms without being an exact duplicate of our world. The physical structure of those other worlds would be indistinguishable from our world, but the intrinsic natures carrying those relations would be different. The most interesting case would be worlds verifying P&~Q. This leads to an interesting metaphysical picture: the relational-structural properties of physics in our world do not necessitate the Q-properties (phenomenal properties), the Q-properties do not supervene logically on the relational-structural properties. However, the relational-structural properties of physics *together* with additional intrinsic properties necessitate the emergence of phenomenal consciousness. This metaphysical picture has been eloquently expressed by Astronomer Sir Arthur Eddington in his work *Space, Time, and Gravitation*: "Physics is the knowledge of structural form, and not knowledge of content. All through the physical world runs that unknown content, which must surely be the stuff of our consciousness" (Eddington 1920, 200). Russell's "neutral monism" was based on similar intuitions: "As regards the world in general, both physical and mental, everything we know of its intrinsic character is derived from the mental side, and almost everything we know of its causal laws is derived from the physical side" (Russell 1927, 402). The structure of Chalmers' argument thus comes finally down to this:

(1) P&~Q is conceivable.
(2) If P&~Q is conceivable, then P&~Q is 1-possible.
(3) If P&~Q is 1-possible, then P&~Q is 2-possible *or* Russellian monism is true.
(4) If P&~Q is 2-possible, materialism is false.
(5) Materialism is false *or* Russellian monism is true.

In the end, Chalmers thinks that traditional physicalism fails because he rejects the idea of a brute and inexplicable emergence of the mental from the physical. Only if there are (proto-)mental properties (intrinsic), nomologically connected to the physical properties (relational) by irreducible emergence laws, can the emergence of consciousness be necessitated in such a way that is no longer mysterious. Galen Strawson has in recent years developed a metaphysical theory based on a similar intuition. He calls it "realistic monism" or "real physicalism" (Strawson 2006). It is based on a venerable argument for panpsychism, the "genetic argument." It rests on a claim about the intelligibility of radical inter-attribute emergence, as opposed to weaker intra-attribute emergence. It mirrors the traditional distinction between a strong inter-actionist substance dualism and a weaker intra-actionist dual-aspect theory or property dualism. In his paper, "Panpsychism" Thomas Nagel had argued that uniform psycho-physical correlations could not account for the emergence of mental properties from the physical components of a system: "Instead, intrinsic properties of the components must be discovered from which the mental properties of the system follow necessarily. This may be unattainable, but if mental phenomena have a causal explanation such properties must exist, and they will not be physical" (Nagel 1979, 187). Emergence cannot be brute in the sense of there being nothing in the emergence base in virtue of which the emergent phenomenon emerges: *ex nihilo nihil fit.* Nothing can give what it does not possess. In order to make sense of the emergence of consciousness, the evolution of the phenomenal mind must be smooth. Consciousness in some form must be present "at the very origin of things" (James 1890, 149). Inter-attribute emergence is to be avoided. Otherwise even a Cartesian soul could mysteriously emerge from a Cartesian physical body. One might as well try to "imagine" the emergence of a concrete physical object from a configuration of abstract mathematical objects. If such inter-attribute emergence is considered intelligible, then one will inevitably end up with a radical Humean view of causal powers in which "any thing may produce any thing" (Treatise, III, xv), and any reasonable prospect for a substantive metaphysical account of the mind-body relation will have vanished. If emergence is construed as an intra-attribute relation, however, then one has to give up the standard physicalist principle that there are no (proto-)mental properties at the ontological base levels. Strawson calls this dogma of contemporary physicalism the "NE principle": "physical stuff is, in itself, in its fundamental nature, something wholly and utterly non-experiential" (Strawson 2006, 11). The intuition that a system that is only structurally defined cannot give rise to qualitative experience is more than just an epistemic problem of cognitive upward opacity that might be overcome by more empirical research; it is a deep metaphysical puzzle. One of the best ways to present this lack of analysis is the "argument from cellular automata" by Gregg Rosenberg (Rosenberg 2004, 14-30). Cellular automata are artificial digital worlds consisting of basic particulars called "cells" in an abstract space. These cells have relational properties connecting them to other cells. Computer modelers define various worlds by giving the cells different properties and then study their dynamics through consecutive computational steps. This is usually done by defining rules that determine which properties a cell will have at a given time as a function of which properties the neighboring cells had at an immediately preceding time. In simple versions of cellular automata the basic particulars have only simple properties like "on" and "off." One can build on these humble beginnings and construct more complicated cellular automata that may ultimately mimic physical properties like spin, charge or mass. The fascinating fact is that despite its rather simple physics the cellular automaton is enormously versatile, in fact a universal Turing machine. Very quickly individual cells join together to build more and more complex structures and patterns that are sustained over many steps of computation. The machine seemingly produces endurants (stable relational patterns) which arise out of a sequence of event-like occurrents (discrete computational states of the system). Quickly these patterns become quite sophisticated, featuring a kind of non-trivial self-replication that is functionally similar to certain structures of living beings, like DNA. For this reason cellular automata are sometimes called "life worlds." The basic facts of cellular automata, the distribution of properties over the grid of cells, necessitate the higher-level structural facts about stable emerging patterns. There is no mysterious inter-attribute emergence involved, even though the emerging patterns exhibit new properties that cannot be attributed to individual cells. Everything happens within one strictly delineated ontological scheme. Standard physicalism assumes that our world is an extremely complex cellular automaton. Thus the following problem arises:

(1) The fundamental facts of cellular automata are defined entirely by the dynamic relations of the cells.
(2) Facts of phenomenal consciousness are intrinsic qualitative facts, which cannot be entirely defined by the dynamic relations in which they enter.
(3) Facts about dynamic relations do not entail (a priori or a posteriori) intrinsic qualitative facts about phenomenal experience.
(4) Thus, the intrinsic qualitative facts about phenomenal experience are not entailed in the facts about cellular automata.

The question is then: if our world is a cellular automaton, how does the phenomenal mind emerge? This is exactly the problem raised by the "genetic argument" for panpsychism. To avoid this notoriously hard problem, one can resort to the denial of the existence of phenomenal experience and become an eliminativist about the phenomenal mind. Strawson is quite outspoken about this strategy: "This particular denial is the strangest thing that has ever happened in the history of human thought, not just the whole history of philosophy. It falls, unfortunately, to philosophy, not religion, to reveal the deepest woo-woo of the human mind" (Strawson 2006, 6). It seems more reasonable to assume that the functional-relational picture of the world assumed in the model of cellular automata is somehow incomplete. This was indeed Russell's intuition. Maybe we are missing something about the intrinsic nature of the physical world in virtue of which (plus the relevant laws) the emergence of conscious mind can be explained. This is, of course, just a conjecture, but it is certainly a *possibility* that knowledge of the intrinsic properties of matter would help to overcome the puzzle of inter-attribute emergence. In Russell's words: "The physical world is only known as regards certain abstract features of its space-time structure – features which, because of their abstractness, do not suffice to show whether the world is, or is not, different in intrinsic character from the world of mind" (Russell 1948, 240). To use Whitehead's term, the modern notion of matter presents us with "vacuous" entities whose intrinsic nature is unknown. Inspired by Humean arguments, Peter Unger has recently presented a visually compelling picture of this problem. Let us define two worlds in purely structural terms, not assuming any intrinsic qualitative properties. The first world is a classical Newtonian world of particles moving about in empty space according to the laws of physics. Call this the "particulate world." In the second world there is a continuous material plenum (a continuous field of matter) in which there are little perfectly empty spaces, or absolute vacua, or simply "bubbles." Call this the "plenumate world." Now let us assume that the two worlds stand in an isomorphic relation in such a way that for each particle in the particulate world there is a corresponding bubble in the plenumate world (in the same location, governed by the same laws). It is Unger's contention that these two worlds are functionally equivalent (Unger 2006, 21-31). A functional-relational description would be unable to capture the differences between these two worlds. To put it in different terms: If we were to construct cellular automata to model these worlds, only one would be needed to model both. Maybe this picture relies too much on visual imaginability, but it nevertheless captures the intuition that the abstractness of the relational structure really "abstracts away" from something that needs to be added to complete the metaphysical picture. The genetic argument for panpsychism claims that we have good reason to believe that this "something," which is being abstracted away from, better be something (proto-)mental, since otherwise the emergence of the phenomenal mind remains mysterious. But the problem of emergence is not the only reason that could motivate a critique of the functional-relational picture. A similar line of argument was developed long before contemporary philosophy of mind. Modern philosophy was quite critical of Descartes' notion of a material substance, a notion that still governs the scientific world-view to this day.

3. Intrinsic properties and the metaphysics of natural individuals

The metaphysical question is whether something as abstract as a formal system of spatio-temporal relations is sufficient to account for the concrete substantiality or thinghood of a natural individual. A structure must be a structure *of* something. Robert Adams recently presented a line of thought similar to Unger's. He claims that we cannot imagine a shape without some chromatic property. The formal entity needs a "filling." And from there he goes on to argue: "We may conjecture that the reality of a substance must include something intrinsic and *qualitative* over and above any formal or structural features it may possess" (Adams 2007, 40). But early modern thought had expelled all qualitative, non-structural qualities from material bodies and had placed them entirely inside the mind. A view that Whitehead has characterized brilliantly: "The poets are entirely mistaken. They should address their lyrics to themselves, and should turn them into odes of self-congratulation on the excellency of the human mind. Nature is a dull affair, soundless, scentless, colourless; merely the hurrying of material, endlessly, meaninglessly" (Whitehead 1925, 80). Assuming a "real physicalism" (in the sense of Strawson) we cannot take refuge in Cartesian dualism or mysterious emergence. Human persons are natural individuals and material objects. But if it is from the experiential qualities that human persons derive their positive non-formal content as *substances*, could we then not reasonably ask if humans are but one special case of natural individuals and that natural substances *as such* require such positive content. We would thus be siding with Whitehead in claiming that some sort of qualitative intrinsic content is constitutive of nature at large. A quick survey of some topics in early modern philosophy will prove helpful in exploring this issue.

In a brilliant piece of skeptical reasoning, Hume challenges the notion of a Cartesian material substance. He claims that upon the removal of sensible qualities from the rank of independent existences, we are merely reduced to primary qualities like figure, motion and cohesion. This process, instead of aiding in explaining the operations of external objects, utterly annihilates all these objects, resulting in the most extravagant skepticism concerning them. And he goes on: "If colours, sounds, tastes, and smells be merely perceptions, nothing we can conceive is possest of a real, continued, and independent existence; not even motion, extension and solidity, which are the primary qualities chiefly insisted on" (Hume 1739, IV, iv, 512f.).

Indeed, Descartes claims that spatial extension is the essence of corporeal substance, and nothing else contributes to it. For Descartes the very nature of a substance is determined by its attributes. He does not construe substances as mere substrata or bare thisness. He argues that the distinction between the notion of a substance and its attributes is merely a distinction of reason not a real distinction: "For there is some difficulty in separating the notion of substance from the notions of thought or extension, which of course differ from substance only in the reason" (Principles 1.63). A material Cartesian substance is thus nothing but modes of extension, i.e., shape, size, and motion in space. The key question for our present purposes is whether or not this notion of a substance as mere extension is ultimately intelligible. What can it mean that there is no real distinction between body and spatial extension? It seems that there has to be *something* that is extended in space, and that something cannot again be just space. Here the problem with relationalism resurfaces again. In his critique of Descartes' notion of a material substance, Leibniz argues that extension can be analyzed in merely relational terms; but then the question about the intrinsic nature of the relata arises. In a letter to de Volder, Leibniz makes the point that extension cannot be conceived in itself. Extension for him, is not a primitive but an analyzable concept; it can be analyzed into plurality, continuity, and coexistence or the existence of parts at one and the same time.[1] But parts of what? As Leibniz argues elsewhere, extension is just a continuous multiplicity of something that is spread out. The nature of the substance that is *being spread out* is not explicated by the concept of extension; on the contrary, it is ontologically prior to the repetitive multiplicity of extension (G IV, 467). Challenging Descartes, Leibniz claims that extension cannot be something absolute, it rather is something relative to what is being expanded: "... *extensionem non esse absolutum, quoddam praedicatum, sed relativum ad id quod extenditur sive diffunditur*" (G IV, 394). The key intuition here is again that extension, due to its formal-relational character, is too abstract to constitute a concrete individual substance in the same way that neither multitude nor number constitute a substance, unless we are ready to specify what it is that is repeated or numbered. Something ontologically prior must be assumed which is continued or diffused. Leibniz gives these intuitive examples: "whiteness in milk, color, ductility and weight in gold, and resistance in matter" (to de Volder, IV 1699, G II, 170). An "extended thing on its own" is impossi-

1 Leibniz to de Volder, IV 1699, G II, 169f. I owe this quote and some other helpful references to early modern philosophy to the unpublished manuscript "Physicalism and Absolutely Intrinsic Properties" by Derek Pereboom. It is available online at consciousness.anu.edu.au/papers/pereboom.doc. Adams 2007 also provides some illuminating references to Leibniz, Hume, Locke and Kant. A new English translation of the Leibniz - de Volder correspondence has been completed by Paul Lodge and will be published in 2009. The drafts are available at: http://users.ox.ac.uk/~mans1095/devolder.htm

ble, it could not even be extended. The relational properties of substances must have an underlying foundation of intrinsic properties. Famously, Leibniz claimed that there is no denomination so extrinsic that it does not have something intrinsic as its foundation (to de Volder, IV 1702, G II, 240). If this is correct, then we need ultimate intrinsic properties that carry the entire net of functional-relational properties in the world. This intuition resurfaces in contemporary debates. In his paper "Pattern and Being", John Haugeland assumes the traditional view that a substance needs properties which it has regardless of anything else. He then considers the ontological status of the pieces in a chess game – say a rook or a pawn – and claims that their very nature is determined entirely by how they move about in the chess game in relation to other pieces. He then concludes: "No rook is a substance. ... Nothing about a rook is determinate, not even its 'rookness,' apart from its participation in a chess game" (Haugeland 1993, 63). The situation is even more complicated. The formal definition of a type in a chess game is circular. The nature of each type is completely determined by the set of allowable moves it makes within the game as a whole. The chess game as a whole, however, is defined by the interdependent set of types which play functional roles in it. Each part of the game presupposes the existence of the whole game, and the game presupposes the existence of its parts. Why isn't this circularity of chess categories vicious (cf. Rosenberg 2004, 234)? How can chess games actually and concretely exist? Rosenberg claims that there must be something distinct from the formal structure that actually grounds the game in concrete reality. In the case of a chess game we have physically distinct objects that serve as stand-ins or realizers of the relevant types, thus allowing for the existence of concrete tokens of those types. Of course, there is much more to consider here, like the concrete chess board or the physical position of the players in space. Without such "carriers" of the formal structure, the game would remain too incomplete and abstract to exist concretely. Rosenberg extends this thought to other, more complex, conceptual systems such as those constructed by scientific theories. They too are abstract and circularly defined. Consider cellular automata in computer science again. Each cell is defined by its role in the entire system, and the entire system is defined by the cells. Cellular automata may exist as computational systems because there is something external to the formal system that realizes or carries it. The physical states of the hardware are the carriers of the cellular automata. Biology as an abstract conceptual system is carried by the mechanics of molecular biochemistry, psychology by the dynamical properties of the neural system, economics by the needs and desires of individuals. The crucial question is however: What carries the most basic physical level? Physics presents us a world of interdependently defined functional roles. Are there any properties that can give this circularly defined conceptual system a foothold in concrete reality? This is a puzzling question. Let us call it the "ultimate carrier problem." It is very similar to the question Leibniz raised with regard to Descartes' notion of matter. Not surprisingly, Rosenberg makes a similar move. He calls properties that are functionally defined within a system "intrinsic to a system" (Rosenberg 2004, 237). In order to avoid an infinite regress of ever more fine-grained systems, where each lower structure is serving as the carrier of the next higher one, a stopper is needed. Also, there are good scientific reasons to assume that nature has a lower size limit (Planck size scale). Only a property that is intrinsic *tout court* and not relative to a system could bring this about and serve as an ultimate carrier. Are there properties that are not intrinsic to any system, but at least partly intrinsic to themselves? The only candidates we know of are phenomenal qualities. One cannot understand the nature of phenomenal qualities by knowledge of their contextual relations alone. This radical intrinsicness is the very nature of phenomenal qualia. *Whatever grounds the structural-relational properties of the world must have this radical intrinsicness.* It might well be that *our own consciousness* is the closest analogue we get to this underlying reality.

A viable metaphysical alternative seems to be radical relationalism. Relationalism differs from functionalism exactly by getting rid of all realizers. According to relationalism, there is only the relational reality consisting of relations and nodes defined by their place in the overall relational system (cf. Dipert 1997). Nature is like a mathematical graph. There are no non-relational properties. But since we know that our own intrinsic qualitative conscious phenomenal states exist, we already know that relationalism as a universal metaphysical doctrine cannot be true. There is more to the world than relations and point-like relata without intrinsic properties. Combining these two insights opens up a rationally well-grounded road to the acceptance of pan(proto-)psychism. This line of thought is often characterized as the "argument for panpsychism from intrinsic natures." It is at this point that the arguments for the irreducibility of qualia in the philosophy of mind and arguments for the intrinsic properties of natural individuals from general metaphysics start to point in the same direction.

This insight may be surprising to contemporary readers, but was well-known by many modern philosophers. For Kant it was obvious that the only absolutely intrinsic properties we can conceive of are taken from the mental realm. It is worthwhile to quote him at length: "It is quite otherwise with a *substantia phaenomenon* in space; its inner determinations are nothing but relations, and it itself is entirely made up of mere relations. We are acquainted with substance in space only through forces which are active in this and that space, either bringing other objects to it (attraction), or preventing them penetrating into it (repulsion and impenetrability). We are not acquainted with any other properties constituting the concept of the substance which appears in space and which we call matter. As object of pure understanding, on the other hand, every substance must have inner determinations and powers which pertain to its inner reality. But what inner accidents can I entertain in thought, save only those which my inner sense presents to me? They must be something which is either itself a thinking or analogous to thinking" (CPR B321, transl. Norman Kemp Smith).

4. The possibility of pan(proto-)psychism reconsidered

By "analogous to thinking" Kant probably meant something like protomental properties that are in some relevant aspects similar to known mental properties but then also quite different from the highly developed mental properties of humans. But how could possibly all relational properties of material objects be grounded in intrinsic properties? And what does "grounding" mean in this context? If grounding means that the intrinsic properties are the constitution base of the relational extrinsic properties, then we have a metaphysical system in which certain fundamental entities with absolutely intrinsic properties constitute all the remainder of reality – a view, that Leibniz famously argued for in his Monadology. The key was to replace "mutual causal influence" with "mutual information" (Seager 2006, 4). The whole of physical space was constructed by giving each monad a spatial viewpoint from which space was constituted as *experienced space*. This seems too daring for most, and is considered by many as a position completely at odds with contemporary science. The latter claim may not be entirely accurate, however. Physicist David Bohm made a somewhat similar claim by introducing the quantum potential in order to provide an ontology for quantum mechanics. In the Bohmian interpretation, a single electron (say) passing to the lower slit in the two-slit experiment receives information by the quantum potential about the state of the entire system, including whether the upper slit is open or shut. It acts according to this information. Bohm talks about "active information", thus introducing mental representation and mental causation (causally efficacious mental content) at the most basic level of the universe (Bohm 1990). But it seems that electron is just mirroring the space around it, using the information presented in the quantum potential. Such mirroring would not be enough to actually *constitute* the spatial relations. This could only be the case if by being experienced in a certain way, spatial relations would come to exist in a certain way. This idea is clearly more related to another interpretation of quantum mechanics – the idea that reality becomes concrete and determined only relative to an experiencing observer. Here the experiencing observer becomes, at least partially, constitutive of reality. In his *Mindful Universe* Henry Stapp advances an interpretation of quantum mechanics according to which entities are ultimately bipolar – material and mental. The Schrödinger equation describes the deterministic and material aspect, the indeterministic collapse relates to the perspectival and mental aspect of reality. This mental aspect does not arise miraculously from the material aspect, rather it is a fundamental feature of reality (Stapp 2007). But without doubt, Leibniz' view goes considerably further; his idealism grounds all physical objects in mental states of the monads, thus rendering the physical world but a "well-founded phenomenon."

This idealistic reading of the absolute carrier problem seems unnecessary. It is not required to claim that all relational facts are mere phenomena. It would suffice to show that relational properties need to be completed and "filled" by absolutely intrinsic properties in order to gain a foothold in concrete reality. To distinguish this solution to the problem of ultimate realizers from Leibniz's idealism, one should characterize it as a dual aspect theory; the relational properties account for the structural form, but the absolutely intrinsic properties account for the ultimate realizers of the relational structure. One might even think of some kind of "hylomorphism" of the relational and the intrinsic. Both aspects together constitute a concrete natural individual. Thus the relational and the intrinsic aspects of reality have basic ontological status, without one having clear priority over the other. Alternatively, a neutral monism can also serve as the metaphysical framework. In this case, the basic properties of the world are neither physical nor phenomenal, but the phenomenal and the physical are constructed out of them. "From their intrinsic natures in combination, the phenomenal is constructed; and from their extrinsic relations, the physical is constructed" (Chalmers 1996, 156). But in any case, this ontology implies that perspectival representation is a fundamental feature of the world. Each natural individual has some representational perspective on the world. Mere conglomerates do not (as such) have this perspectival unity, only the simples that constitute them may again be unified in this way. The classic distinction between true substances and mere conglomerates can thus be explained in this metaphysical picture. Mental or proto-mental intrinsic properties can thus play an important role in solving the "special composition problem" for substances.

In recent philosophy of mind there has been a discussion whether consciousness is grounded in intentionality, or intentionality is grounded in consciousness. Reductive representationalists argue that phenomenal properties are equivalent to representational properties that can be completely characterized in non-phenomenal terms. If this is true, then built-in representational perspectives will not solve the absolute carrier problem because they are again merely structural. In order to solve the absolute carrier problem we need carriers that are absolutely intrinsic. Again, the only candidates for properties of this kind *we know of* are phenomenal, experiential mental properties. It is this fact which lends substantial support for some kind of panpsychism.

It was the purpose of this paper to show that this move is not only motivated by non-reductionist arguments in the philosophy of mind but equally by the argument for intrinsic natures in the metaphysics of concrete particulars. It is this mutual support of independently well-motivated reasons that makes the overall argument compelling. This complex edifice of metaphysical assertions does not come without a price, however. Thomas Nagel has famously remarked that "panpsychism has the faintly sickening odor of something put together in a metaphysical laboratory" (Nagel 1986, 49). The position is initially counter-intuitive, and only as a result of lengthy argumentation is it gaining some plausibility. There are, of course, "simpler" alternatives. Assuming so far unknown absolutely intrinsic properties of matter is certainly a possibility. Russell sometimes preferred skepticism: "The only legitimate attitude about the physical world seems to be one of complete agnosticism as regards all but its mathematical properties" (Russell 1927, 270). This sounds reasonable indeed, and a certain amount of skepticism with regard to absolutely intrinsic properties is well-advised. But that leaves us with a significant gap in our understanding. On the other hand, there remains the intuitive force of the genetic argument: "we ought … to try every possible mode of conceiving of consciousness so that it may not appear equivalent to the irruption into the universe of a new nature non-existent to then" (James 1890, 148).

References

Adams, R. (2007) 'Idealism Vindicated' in van Inwagen, P. and Zimmerman, D. (eds.), *Persons. Human and Divine,* Oxford: Oxford University Press, 35-54.

Bohm, D. (1990) 'A New Theory of the Relationship of Mind and Matter', *Philosophical Psychology* 3, 271-286.

Chalmers, D. (1996) *The Conscious Mind. In Search of a Fundamental Theory,* Oxford: Oxford University Press.

Chalmers, D. (2002) 'Consciousness and its Place in Nature' in: Chalmers, D. (ed.), *Philosophy of Mind*, New York: Oxford University Press, 247-272.

Chalmers, D. (2009) 'The Two-Dimensional Argument Against Materialism' in: McLaughlin, B., Beckermann, A., and Walter, S. (eds.), *The Oxford Handbook of the Philosophy of Mind*, Oxford: Oxford University Press,
available at http://consc.net/papers/2dargument.html.

Descartes, R. (1644) *Principles of Philosophy*, Translation with Explanatory Notes (Synthese Historical Library) by Miller, R.P (1991), Dordrecht: Kluwer.

Dipert, R. (1997) 'The mathematical structure of the world: the world as graph', *Journal of Philosophy* 94, 329-358.

Eddington, A. (1920) *Space, Time and Gravitation*, Cambridge: Cambridge University Press.

Haugeland, J. (1993) 'Pattern and Being' in: Dahlboom, B. (ed.), *Dennett and His Critics*, Cambridge: Blackwell, 53-69.

Hume, D. (1739) *A Treatise of Human Nature*. New Edition. London: 1886.

James, W. (1890) *The Principles of Psychology*, *vol. 1*, New York: Henry Holt. Reprinted 1950: New York: Dover Publications.

Nagel, T. (1979): 'Panpsychism' in: Nagel, T., *Mortal Questions*. Cambridge: Cambridge University Press, 181-195.

Nagel, T. (1986) *The View from Nowhere*. Oxford: Oxford University Press.

Newman, M. (1928) 'Mr. Russell's Causal Theory of Perception', *Mind* 37, 137-148.

Putnam, H. (1980) 'Models and Reality', *The Journal of Symbolic Logic* 45, 464-482.

Rosenberg, G. (2004) *A Place for Consciousness. Probing the Deep Structure of the Natural World*. Oxford: Oxford University Press.

Russell, B. (1927) *The Analysis of Matter*. London: Routledge.

Russell, B. (1948) *Human Knowledge. Its Scope and its Limits*. London: Routledge.

Seager, W. (2006) 'Rosenberg, Reducibility and Consciousness', *Psyche* 12, 1-15.

Stapp, H. (2007) *Mindful Universe: Quantum Mechanics and the Participating Observer*. New York: Springer.

Strawson, G. (2006) 'Realistic Monism' in: Strawson, G. (ed.). *Consciousness and its Place in Nature. Does Physicalism entail Panpsychism?*, Exeter: Imprint Academic, 3-31.

Whitehead, A. N. (1925) *Science and the Modern World*. Reprinted: New York 1997: Simon and Schuster.

LIST OF AUTHORS

LYNNE RUDDER BAKER
Department of Philosophy, University of Massachusetts Amherst (USA)

GODEHARD BRÜNTRUP
Hochschule für Philosophie, München (GER)

CODY GILMORE
Department of Philosophy, University of California at Davis (USA)

FRANK HOFMANN
Philosophisches Seminar, Universität Tübingen (GER)

LUDGER HONNEFELDER
Otto Warburg Senior Research Professur, Theologische Fakultät der Humboldt-Universität zu Berlin (GER)

LUDGER JANSEN
Institut für Philosophie, Universität Rostock (GER)

SIMO KNUUTTILA
Department of Systematic Theology, University of Helsinki (FIN)

E. JONATHAN LOWE
Department of Philosophy, Durham University (GB)

UWE MEIXNER
Institut für Philosophie, Universität Regensburg (GER)

STEPHEN MUMFORD
Department of Philosophy, The University of Nottingham (GB)

EDMUND RUNGGALDIER
Guardini Professur für Religionsphilosophie und katholische Weltanschauung, Theologische Fakultät der Humboldt-Universität zu Berlin (GER)

JAN SZAIF
Department of Philosophy, University of California at Davis (USA)